Ricardo Cunha Lima

Limnetic Bivalves

Ricardo Cunha Lima

Limnetic Bivalves

A strategy for the conservation of the genus Diplodon

ScienciaScripts

Imprint

Cover image: www.ingimage.com

This book is a translation from the original published under ISBN 978-3-330-75901-5.

Publisher:
Sciencia Scripts
is a trademark of
Dodo Books Indian Ocean Ltd. and OmniScriptum S.R.L publishing group

120 High Road, East Finchley, London, N2 9ED, United Kingdom
Str. Armeneasca 28/1, office 1, Chisinau MD-2012, Republic of Moldova, Europe
Printed at: see last page
ISBN: 978-620-8-30637-3

Summary

"When talking about evolution,

Today, it's never good,

Tomorrow will always be better"

Richard Dawkins

Thanks

To Prof. Dr. Wagner Eustàquio Paiva Avelar for his insistence, patience and collaboration.

Prof. Dr. Sonia Helena Sipauba and Prof. Dr. Joao Batista Kochenborger Fernandes of CAUNESP/UNESP, Jaboticabal, for providing the microalgae strains and fish, respectively, needed for the research.

To Prof. Dr. Luiz Ricardo Lopes Simone of the Zoology Museum (MZUSP), for his supervision, conversation, exchange of ideas and support.

To all the staff of the Department of Biology and the Secretariat of the Comparative Biology Program who directly and indirectly helped me, especially Prof. Dr. Fernando Mantelatto, for his suggestions and advice.

The Coordination for the Improvement of Higher Education Personnel (CAPES) and the São Paulo State Research Foundation (FAPESP) for the scholarship and financial aid granted.

Summary

The lymnic bivalves, or non-natives of the order Unionoida, represent the largest group of freshwater bivalves, with six families, 181 genera and 800 species. They are found on all continents and currently represent the group of freshwater animals most at risk of extinction. The most dramatic cause of the decline and extinction of dulcicolous bivalves is the modification and destruction of their habitat, global climate change and the introduction of exotic (non-native) aquatic animals.

This taxon is the only member of the class Bivalvia recognized for having a parasitic larval stage in its life cycle, which involves an obligatory relationship with a vertebrate host, usually a fish, and a highly modified larva, the gloquid or lasid. This feature of the life cycle is a key component of any conservation plan for limnetic bivalves.

In order to propagate the endangered species, the aim was to develop viable techniques for obtaining juvenile forms in the laboratory by artificially cultivating the gloquidia (*in vitro*).

Using the methodology employed, it was possible to obtain juveniles of more than 30 days of the species *D. expansus*, *D. rotundus gratus* and *D. martensi*. In addition to these results, this study developed a new culture medium for gloquid larvae, based on a lyophilized fish extract.

Introduction

The Bivalvia, currently under the Diassoma group, are molluscs without a defined cephalic region, with a simple foot attached to the visceral mass, two pairs of gills, and each individual has two valves, composed of calcium carbonate, around the body. The class includes around 20,000 living species found in marine and freshwater environments, most of which feed on microphytes or suspensors (Brusca and Brusca, 2007).

Although there are some inconsistencies between the Bivalvia classification schemes, most of the arrangements are consistent with the taxon, divided into two subtaxons: Protobranchia and Autobranchia (=Isofilibranchia + Pteriomorpha + Anomalodesmata + Heterodonta + Palaeoheterodonta). The current Palaeoheterodonta include only one marine genus, Neotrigonia, and the superfamily Unionacea (Giribet and Wheeler, 2002; Graf and Cummings, 2006). The division between the Autobranchia and the inclusion of the Unionoida among the Palaeoheterodonta was traditionally based on hinge morphology (Thiele, 1934), umbonal sculpture (Modell, 1964), and ctenidium morphology (Newell, 1969).

Among the Palaeoheterodonta there is a very peculiar taxon, the limnic bivalves, or naiads, of the order Unionoida, representing the largest group of freshwater bivalves. This diverse assemblage is divided into six families, 181 genera and 800 species (Graf and Cummings, 2006). These families are found on all continents except Antarctica (Haas, 1969), although a Paleozoic fossil of a limnetic bivalve (Anthracosioidea) has been found on this continent (Bradshaw, 1984); and they currently represent the group of freshwater animals most at risk of extinction (Bogan, 2008).

This radiation is unique in the class Bivalvia, forming a probably monophyletic taxon, supported by the restriction to fresh water, ovoviviparity, and a parasitic larval stage that must infest an appropriate host, usually a fish, to complete its metamorphosis (Boss, 1982; Kat, 1984; Schneider, 2001; Wachtler *et al.*, 2001). There are exceptions, both in incubation, with the genus Rhipidodonta presenting a larva with direct development (Bonetto, 1962), and in parasitism, with the species *Lampsilis cardium* parasitizing a species of salamander (Watters and O'dee, 1998).

Based on various schools of malacological taxonomy (Simpson, 1900, 1914; Ortmann, 1910, 1911, 1912, 1921; Frierson, 1927; Modell, 1942; Morrison, 1956, 1973; McMichael and Hiscock, 1958; Pain and Woodward, 1961; Parodiz and Bonetto, 1963; Haas, 1969; Heard and Guckert, 1971; Davis and Fuller, 1981; Boss, 1982; Korniushin, 1998) we can state that the Unionoida are represented by six families: Margaritiferidae, Unionidae, Hyriidae, Iridinidae, Mycetopodidae and Etheriidae.

According to Parodiz and Bonetto (1963), the first three of these families are grouped under the superfamily Unionoidea, and the last three belong to the superfamily Etherioidea. These superfamilies were grouped only by the type of larva they have: the Unionoidea have a gloquid-like larva and the Etherioidea have a lasid-like larva (Parodiz and Bonetto, 1963).

Graf (2000) tested the monophyly of the Etherioidea *sensu* Parodiz and Bonetto (1963) by including the Hyriidae among the Unionoida, coding 38 characters of the shell and soft tissue anatomy, which made it possible to infer that the Unionoidea taxon is not monophyletic and that the Hyriidae group is part of a natural taxon when included among the Etherioidea. Graf and O'Foighil (2000) hypothesized a similar phylogeny using molecular analyses. Therefore, Etherioidea *sensu* Graf (2000) and Graf and O'Foighil (2000) [= (Hyriidae + (Iridinidae + Etheriidae))] is a synonymy of Mutelidae of Ortmann (1912, 1921), Thiele (1934) and McMichael and Hiscock (1958).

Further reinforcing this separation is the zoogeography of these two taxa, with the Unionoidea in the northern hemisphere and the Etherioidea showing a Gondwana distribution (Ortmann, 1921).

Distribution

Research into limnetic bivalves has revealed some macro-evolutionary patterns. The Unioniformes are a cosmopolitan, ancient and strictly continental group. Given that these patterns are accurate and that geographic separation contributes to the diversification of organisms (Mayr, 1963), it is expected that the biogeography, evolution of characters and phylogeny of freshwater bivalves should reflect the evolution of continental masses during the Mesozoic and Cenozoic.

Naiads first appeared in the Triassic (Carnian / Rhaetian) and Neogene (Miocene / Pliocene), after the mass extinction at the end of the Permian (Graf, 2007). In contrast, almost all the other bivalve families that originated in the Cretaceous (Aptian / Cenomanian) and Paleogene (Danian / Thanetian) were marine (Newell, 1969). This seems to coincide with the Wilson cycle corresponding to aggregation (with the formation of freshwater fluvial and lacustrine environments) and continental break-up (with the formation of seas and young saltwater oceans). In other words, continental aggregation tends to produce brackish/sweetwater faunas, and the separation of continents can cause marine fauna to evolve (Kondo and Sano, 2009).

Parodiz (1969) made an excellent review of the fossil terrestrial malacofauna of South America, providing a list of species for various localities, and summarizing the nature of the geological records of each part of the continent. The fossil record described by Parodiz supports Pilsbry's (1911) division of the fauna into three groups: the autochthonous species, the species that entered from North America in the late Cretaceous, and the ancient forms with a global distribution and of uncertain origin (e.g. Sphaeriidae).

Most of the recent fossils from Brazilian areas have affinities with African or Old World groups, and those found in southern localities are of North American origin (Parodiz, 1982).

The interpretation of these recent introductions does not pose any problems, but the duality of the origin of various Paleocene records shows an apparent homogeneity, characteristic of the Neotropical fauna, due to the complete isolation of South America during almost the entire Tertiary (Haas, 1969; Parodiz, 1982). Communication with North America was discontinued immediately after the start of the Paleocene, and was not re-established until the end of the Pliocene, by a new connection through Central America with the rise of the Isthmus of Panama (Rosen, 1978).

During most of the Eogene (lower Oligocene) the crystalline shield of South America (called Brasilia by Ihering) was separated from the lands to the south (Patagonian region) by a sea course connecting the newly formed Atlantic Ocean with what was later the Eastern Pacific (Fitzgerald,

2002). The sea dividing line ran obliquely in a northwest to southeast direction, from southern Peru down to the area known as the Pampas, and according to Parodiz (1982), is partly coincident with the current physiographic division of the continent between the tropical and temperate zones. This southern land mass was called Archiplata by Ihering. The scar or suture between Brasilia and Archiplata is visible in the geological structure of the region between the Colorado and Negro rivers in northern Patagonia (Visconti *et al.*, 2003). Most of the Paleocene localities where the fossils of freshwater bivalves are found are in the formations along the Archiplata cordon, but they are completely absent on the opposite side, the southern coast of eastern Brasilia (Parodiz, 1982).

Therefore, the malacofauna that developed in Brasilia and Archiplata is not only different in its composition, but also in its origin. Most of the species from Archiplata in the Paleocene, known from Patagonia to western Bolivia, have similarities, and in some cases are of the same genus as those known in North America from the pre-Triassic period. The most surprising group is the Hyriidae, which first appeared in the Triassic in Pennsylvania and although it became extinct in this region, it suddenly reappeared in the Paleocene in Patagonia (Parodiz and Bonetto, 1963; Parodiz, 1982).

In South America, apart from the great Amazon Basin, with its main origins in the west, and the Orinoco River drainage in Venezuela, both of which present peculiar distributional restrictions for the species, the main hydrographic systems in the central-eastern part of the continent are the São Francisco River, with its sources on the Minas Gerais plateau, and the Paranà - La Plata system, with its various tributaries flowing to the south (Parodiz, 1982).

Figueiredo *et al.* (2009), using new biostratigraphic data, concluded that the Amazon River became a transcontinental river around 11 million years ago and reached its current shape and size during the late Pliocene. Before the late Miocene there was a continental shelf that led the course of the rivers towards the Pacific.

The Paranà and San Francisco rivers have several elements in common in their beginnings, but the abundance and differentiation of genera and species increases enormously towards the Uruguay and La Plata rivers. The Hyriidae and Mycetopodidae families, with well-known ancestors from the Paleocene to the Miocene, have no fossil record of any age in the Pampas region and the lower middle section of the Paranà River contains only a few Holocene fossils (Parodiz, 1982). The Paranà - Paraguay - Uruguay - La Plata system did not exist before the end of the Pleistocene. And around 11,000 years ago, a major fault occurred, correlated with the third phase of the Andes orogeny. The west bank became several meters lower than the east side. A brief ingression of the sea was immediately followed by the effluents of the currents coming from the north, through the fault, forming a new drainage system and, therefore, the new estuary (Martinez and Del Rio, 2002). According to Parodiz (1982), the limnic fauna introduced into this new basin underwent rapid speciation. Consequently, in the Hyriidae family, several taxonomically recognized species have a very close genetic affinity, due to their recent common ancestor, forming groups equivalent to superspecies. In such cases, hybridization is not only possible, but actually occurs (Parodiz, 1973).

The invasion of species from the north, in the lower system of the La Plata River, probably occurred recently, through the connection of the tributaries of the upper Paraguay River with those of the Amazon River. No species of Mycetopodidae are known west of the Andes, in Peru and Chile, an area occupied by the Hyriidae of the genus Diplodon (Parodiz and Bonetto, 1963).

The Hyriidae originated in the Oligocene in Colombia and especially in the Miocene in Ecuador (Parodiz, 1982). The oldest known *Diplodon* are represented by several species from the Triassic in Pennsylvania and Texas. Another fossil group is found in the Paleocene and Eocene in South America (southern Argentina and Chile). All these fossils are generally smaller than most recent species, except for the Chilean forms from the Eocene, which differ very little from the current species *Diplodon patagonicus*. Other Diplodon fossils from the early Tertiary are also known from Colombia to southern Argentina (Parodiz and Bonetto, 1963). Martinez *et al.* (1993) describe *Tacuaremboia*, a Unionoidea in the Tacuarembó Formation in Uruguay, in deposits of Triassic-Jurassic age. Morton and Herbst (2001) describe *Diplodon* in the La Matilde Formation, from the Middle Jurassic, in the province of Santa Cruz in Argentina. The taxonomic diversity of the Unionoidea in the Cretaceous of Montana, USA was analyzed by Scholz and Hartman (2007). Wesselingh *et al.* (2006) mention the occurrence *of Castalia ambigua*, *Callonaia* sp and *D. longulus* in the Purus and Acre rivers, Acre state, Brazil, in the Solimoes Formation, from the late Miocene.

Comparison of the distribution of recent and Tertiary Hyriidae species shows that expansion occurred from west to east (the oldest forms distributed along the Andes) and later, especially from southwest to northeast between northern Patagonia and the rivers of the Paranâ system.

The rivers that cross the Pampas region, tributaries of the Paranâ, have a higher concentration of salts than the eastern tributaries, and according to Bonetto (1961c) and Parodiz and Bonetto (1963), this is a factor that would have restricted the dispersal of the Hyriidae.

Diversity

The superfamily Unionoidea (= Unionidae + Margaritiferidae) is diverse (686 sp) and is widely distributed across the northern hemisphere; the superfamily Etherioidea (= Hyriidae + Etheriidae + Mycetopodidae + Iridinidae) is relatively less specialized (154 sp), and the diversity of the Etheriidae families is generally split between the southern continents.

The sister group of the Unionidae, the Margaritiferidae family, shares a similar Holarctic distribution, but with significantly less diversity (12 sp).

The freshwater bivalve assemblage of the Afrotropical region (85 sp) is made up of three families. There are two dominant families, the Unionidae (41 sp) and the Iridinidae (43 sp), and a single widely distributed species of the Etheriidae, *Etheria elliptica* (Graf and Cummings, 2007).

The Indotropical region is the second most diverse, with 219 species of limnetic bivalves in three families. Only one of these species, *Pseudomulleria dalyi* in India, represents the Etherioidea (Bogan and Hoeh, 2000; Graf and Cummings, 2006). The rest of the Indotropical species represent the Unionoidea and the northern hemisphere.

The largest family of Etheriidae, the Hyriidae (71 sp) is currently only known from Australasia and South America (McMichael and Hiscock, 1958; Simone, 2006; Graf, 2007). The Mycetopodidae (36 sp) and their sister group Iridinidae (43 sp) are known from the Neotropical and Afrotropical regions, respectively.

There are an estimated 43 species of unionids in the Australasian region. The diversity in this region is dominated by the Hyriidae with eight genera and 28 species (Bogan, 2008). The Hyriidae are restricted to Australia, Tasmania, New Zealand, New Guinea and the Solomon Islands.

The South American limnetic bivalve fauna is of Gondwana origin (Graf, 2000), representing three families endemic to fragments of the aforementioned supercontinent. In general, 74 species are recognized from South America (Haas, 1969; Simone, 2006), less than half (43%) of the total diversity of recognized neotropical species. The two dominant families are the Mycetopodidae (32 sp) and the Hyriidae (40 sp), each probably representing a monophyletic radiation from a Gondwana ancestor (Graf, 2007).

Older authors (Say, 1817; Rafinesque, 1820; Hupe, 1857; Adams and Adams, 1858; Pilsbry, 1911) classified the South American forms of the naiads together with the old collective genera, *Unio* and *Anodonta*. To this group they added *Monocondylaea*, *Hyria* and *Castalia*.

Ihering (1890) recognized that there were differences between specimens from the northern and southern hemispheres under the synonymy *of Anodonta*, and that the latter group shared characteristics with African specimens (*Mutela* and *Spatha*). And he calls it *Glabaris* Gray = *Anodontites* Bruguière. However, the author left the other forms under the name *Unio*.

Simpson (1900) separated the group under the synonym *Unio* and used the name *Diplodon* Spix for the specimens from the southern hemisphere, uniting them with the Australian naiads.

Ortmann (1911) corroborated Simpson's hypothesis by indicating anatomical differences between *Diplodon* (as well as *Hyria* and *Castalia*) and the Unionidae of the northern hemisphere, associating them with *Glabaris*, thus forming the Mutelidae (=Etherioidea) family, subdivided into two subfamilies, Hyriinae and Mutelinae.

The most primitive group, Hyriinae (Ortmann, 1921), is found throughout South America, becoming rarer in the northern region (Colombia and Venezuela), not being found in Central America, but also in Australia.

Parodiz and Bonetto (1963) describe that, in South America, the Hyriidae and Mycetopodidae families occupy overlapping areas, although in marginal zones there is only one group or the other. The Mycetopodidae are more restricted in their southern distribution, occupying areas to the north, east and periphery of the Pampas region, and some tributaries that cross this region and flow into the Paranâ River. In the tributaries on the left side of the Paranà River, the Mycetopodidae are more developed than the Hyriidae. In a north-westerly direction they spread as far as Central America and Mexico, where the Hyriidae are absent (Bonetto, 1961c).

The Mycetopodidae, made up of four endemic subfamilies (Mycetopodinae, Monocondylaeinae, Leilinae, Anodontitinae), have recently been hypothesized to be a sister group to the Afro-tropical Iridinidae (Graf and Cummings, 2006; Bogan and Hoeh, 2000). And the endemic tribes of Hyriinae, Prisondontini, Diplodontini, and Castaliini would form a sister clade of Hyridellini, from Australasia, forming the family Hyriidae (Graf and O'Foighil, 2000). The current distribution of the Hyriidae family, according to Simone (2006), can be seen in Annex A.

Simpson (1900, 1914) was the first to propose a classification system for the genus Diplodon, separating the rounded forms (Cyclomya) from the others. However, it was Ortmann (1921), through detailed anatomical studies and a large number of specimens of each type, who managed to separate the genus into groups with well-defined characteristics. However, the same author admits that there are transitional forms between the groups that are difficult to classify.

The authors Ortmann (1921), Bonetto (1960, 1961a, 1961b, 1962, 1964, 1965), Bonetto and Ezcurra (1963), Parodiz and Bonetto (1963) and Mansur (1973) mention that, in addition to the characteristics of the shells, some particularities of the morphology of the body, the position of the marsupium and the description of the glochidium would be very useful for identifying the species.

Anatomy

The mantle is an organ, similar to a leaf, which forms the dorsal wall of the body, and in most molluscs it grows during development in the form of one or two folds containing layers of muscle and hemocellular canals. This external growth creates a space between the mantle fold and the body proper, the mantle cavity or palial. Bivalves have a large mantle cavity, with the foot and visceral mass in the center. The mantle folds follow the edge of the shells laterally, giving rise to the inhalant and exhalant opening, through which water enters and leaves the mantle cavity.

Five mantle fusion categories are recognized in its posterior portion, simplified as Types I-V (Yonge, 1957) (figure 1).

Type I is the simplest condition. In this type, the posterior mantle is (in general) unfused, and the diaphragm separating the infrabranchial from the suprabranchial chamber is grossly incomplete. The division of the mantle cavity is functionally achieved via pallial grooves.

In Type II there is no fusion of the mantle between the inhalant and exhalant openings, but the ctenidia are fused to the adjacent mantle along its length, forming a slightly incomplete diaphragm. The inner folds of the mantle are fused by a small extension, dorsal to the exhalant opening, and then re-open to form a third opening, the supraanal opening.

In Type III, the inner folds of the mantle are fused between the inhaling and exhaling openings. In Type IV, in addition to the fusion between the openings, there is also a fusion of the inner folds of the dorsal mantle of the exhalant opening, often forming an exhalant siphon. The main distinction between types III and IV is that, in the former, the inner folds are not joined independently of the outer folds of the mantle above the exhalant opening. In certain genera with type IV palpal fusion, there is often a small association of the inner ventral folds with the inhalant opening. The presence of these small appendages seems to vary between species, but they are never associated with the palial sinus.

Type V is characterized by retractable inhalant and exhalant siphons and a palial sinus. The complete range of opening morphology leads to the formation of Yonge's Type A siphons (1957, 1982).

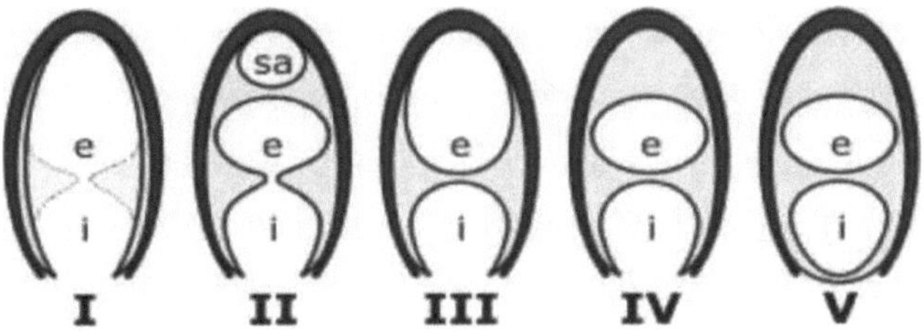

Figure 1. Diagram of the five categories of mantle fusion in its posterior portion. Legend: i, inhalant opening; e, exhalant opening; sa, supranal opening.

Water enters the palial cavity through the inhalant openings and/or siphons, bathes the gills, reaches the suprabranchial cavity and exits through the exhalant openings and/or siphons. Along this route, the suspended material that enters the palial cavity is retained, gas exchange takes place and the products of excretion, defecation and sexual excretion are eliminated.

The cylinders of the ctenidia generate the water current, and the increase in efficiency for food intake was achieved through various modifications to the ctenidia. The main modification was the conversion of the original small, triangular plates into V-shaped filaments with extensions on both sides. The lamella of this V-shaped filament that is attached to the central axis of the ctenidium is called the descending lamella; the lamella forming the other half of the V is the ascending lamella. Normally, the ascending lamella is connected by ciliary contacts or tissue junctions to the roof of the mantle, or to the visceral mass. Taken together, the two V-shaped filaments, with their double row of leaflets, form a W-shaped structure when observed in cross-section.

As the water passes through the interfilamentary spaces, the suspended particles present in the water are retained on the surface of the filaments. Once on the surface, the particles are moved towards the alimentary canal at the free edge of the ctenidium, and then anteriorly to the labial palps. The palps select the material, passing it on to the mouth. Rejected particles fall from the gills or palps into the mantle cavity as pseudo faeces.

The natural system of naiads expresses a tendency towards the development and specialization of three main anatomical features, which in turn are connected to certain functions. These are: the separation of an original simple gill chamber into two chambers, gill and suprabranchial, by the diaphragm; the restriction of the inhalant and exhalant openings to defined parts of the mantle edge, with a tendency to form siphons; and the development of the gills into organs for carrying eggs and larvae (marsupium), and the specialization and adaptation of the gills for this purpose (Ortmann, 1912, 1921).

In the development of the diaphragm, three types can be distinguished. The most primitive state is represented by Margaritiferidae, where the diaphragm is formed by the joint growth of the inner lamella of the inner demibranch on both sides of the body, and the fusion of the outer lamella of the outer demibranch with the mantle. But here the diaphragm is still incomplete, so that the posterior end of the outer lamella of the outer demibranch remains free, not connecting with the mantle at its posterior margin (figure 2A). In all other cases, the diaphragm is complete and extends to or near the posterior margin of the mantle, where it separates the inhalant from the exhalant opening. Two types are recognized, Unionidae (North America and Asia) with the diaphragm formed by the gills alone, and Etherioidea (South America and Africa) with the diaphragm formed anteriorly by the gills, but posteriorly by the joining of the mantle itself.

The edges of the mantle are originally free all around. But a tendency develops to give rise to two distinct openings, the siphons (Yonge, 1957). The most primitive condition is found in Margaritiferidae, where the two openings are distinguished only by the development of papillae.

But the general tendency among Unionidae is not only to join the edges of the mantle by the diaphragm, separating the inhaling and exhaling apertures. But also to limit above the exhalant opening, by joining the edges of the mantle, forming a supranal opening, which only in rare cases becomes closed (figure 2B). The inhalant opening in the Unionidae is never previously defined by a joint growth of the mantle margins (Ortmann, 1911, 1912).

In the other superfamily, Etherioidea, the two apertures are always separated from each other by a complete diaphragm, formed by the mantle (Ortmann,

1921). In some cases the edges of the mantle are free from the rest (e.g. *Anodontites*) (figure 2C). In other cases the exhalant opening is defined above by a junction of the mantle margins (e.g. *Diplodon*), but in this group a supranal opening is never formed (figure 2D). Furthermore, an additional developmental step was found in this group, which consists of the anterior demarcation of the inhalant opening by the joint growth of the mantle edges (e.g. *Castalia*) (figure 2E).

The formation of the marsupium offers great variety. It is difficult to say which is related to the primitive condition, but according to Ortmann (1911), originally all four gills served as a receptacle for the eggs.

Out of this original condition more advanced structures have developed, which generally exhibit the tendency to locate the marsupium in certain gills or parts of the gills, and to divide the interlamellar space into compartments (ovisacs and aquifer tubes). In the different groups, this purpose was achieved in different ways and to different degrees (Ortmann, 1921).

In Margaritiferidae, the gills do not have partitions and apparently ovisacs and aquifer tubes are not found. Consequently, they are primitive in this respect.

In the most primitive Unionidae (subfamily Unioninae), all four gills still serve as a marsupium and have therefore preserved their original condition, or only the outer demibranch serves this purpose. In these cases, the interlamellar space is divided by septa into irregular compartments, arranged vertically to the edge of the gills and parallel to the gill filaments. The same fundamental characters, restriction of the marsupium to the external demibranchs and development of ovisacs and aquifer tubes, are found in the more highly developed Unionidae (subfamilies Anodontinae and Lampsilinae), but here the specializations found may be mainly linked to adaptations to a prolonged reproductive period, and to a peculiar way of releasing the glochidia (Graf and O'Foighil, 2000).

In the Afro-South American group of naiads (Etherioidea), bivalves have their gills restricted to the internal demibranchs. Two main types can be distinguished between them, according to their internal differentiation from the marsupial gill. In one case, the marsupial part of the internal demibranch is somewhat restricted to the median portion, and rather incomplete intercommunicating aquifer tubes are formed by interlamellar connections; in the other case, strong and complete septa form isolated aquifer tubes. In none of these cases is the structure of the marsupial gill identical to that of the Unionidae, although the septa and aquifer tubes show some analogy. Ortmann (1911) considers this to be a case of structural convergence.

North American unioniforms differ from the African and South American types in the anterior junction of the internal demibranchs and the shape of the palps. The most significant difference between Margaritiferidae and the other forms of North American unionids lies in the shape of the inhalant and exhalant openings. In the former, the inhalant and exhalant openings form a large, uninterrupted opening.

Anatomically, North American unioniforms are characterized by a noticeable decrease in the width of the inner demibranch towards its anterior end; this is more or less in front of the anterior end of the outer demibranch, and is attached, at greater or lesser distances, to the ascending portion of the mantle attachment line, but is always distinctly separated from the posterior end of the palpus, usually by a considerable interval. The shape of the palpus is not so well marked, but they are more or less sickle-shaped, with the tip joined posteriorly. The posterior margin of the two palps can be more or less connected or almost free (figure 2B).

The Asian forms resemble the North American ones in terms of the inhalant and exhalant openings, the diaphragm, the palps and demibranchs. The only difference is the separation of the anal and supranal openings, which is slightly longer, at the junction of the inner lamella of the internal demibranch and the visceral mass. Externally, the main conchological feature is the radial sculptures.

African naiads are distinguished from freshwater bivalves from North America and Asia by the junction of the ascending lamella of the internal demibranch with the ascending portion of the mantle adhesion line, which is adhered along its entire length, making contact between its anterior end and the posterior end of the palps (figure 2F).

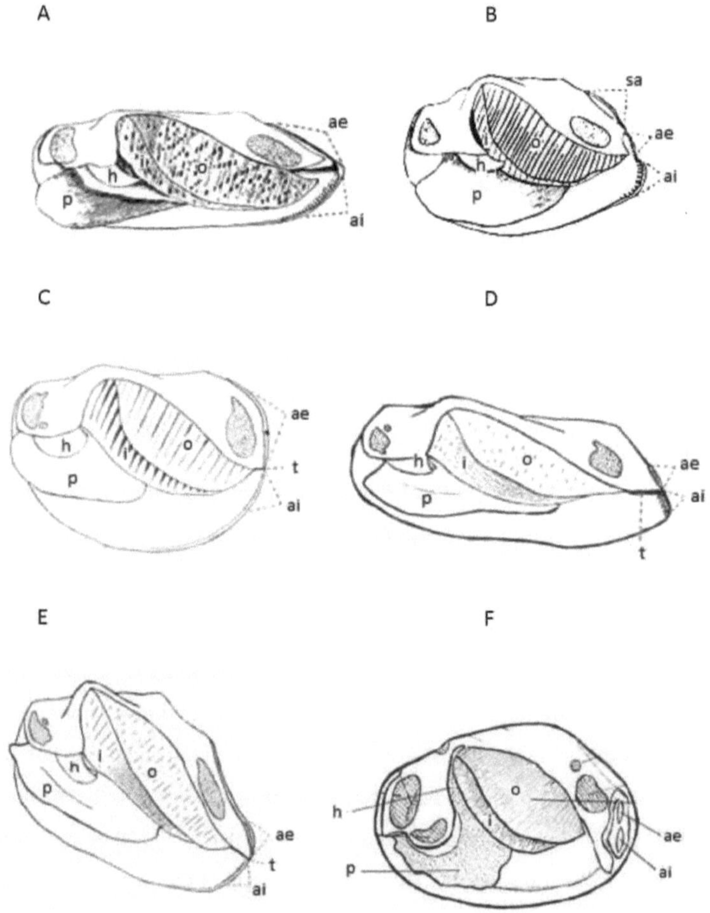

Figure 2. Unioniform anatomy. A, Margaritifera margaritifera; B, Fusconaia rubiginosa; C, Anodontites patagonicus; D, Diplodon trifidus; E, Castalina nehringi; F, Spatha kamerunensis. Legend: ai, inhalant aperture; ae, exhalant aperture; sa, supranal aperture; p, foot; o, external demibranch; i, internal demibranch; h, palps; t, diaphragm. Drawings modified from Ortmann (1912b, 1921).

The palps are wide and short, with a very insignificant tip at the back, and not entirely sickle-shaped; their posterior margins are not connected.

The inhalant and exhalant openings are well separated by a bridge, formed by firmly joining the edges of the mantle, subsequently forming a diaphragm, with the gills not reaching the posterior margin of the mantle.

The exhalant opening is closed above by joining the inner edge of the mantle without leaving a supranal opening.

The septa of the well-developed aquifer tubes are far apart and strong. In females, only the internal demibranchs are used as a marsupium, with the septa widening.

The South American naiads of the subfamily Hyriinae are divided by the characteristics of the shell, but also by the characteristics of the structure of the gill and shell, into three tribes Prisodontini, Castaliini and Diplodontini, including eight genera.

The Prisodontini tribe is characterized by a subromboidal shell, bialate or winged only at the back, but always with a large posterior expansion. Radial umbonal sculpture, very strong, with conspicuous coalescence of the vertical ribs; rarely the sculpture may be non-conspicuous. Well-marked posterior sulcus. The inhalant opening is not completely closed at the front, but continues with the podal opening. Glochidia triangular (isosceles), with shorter and less curved teeth than Diplodontini, ending in 2-3 sharp points. Genera Prisodon, Paxyodon and Triplodon.

The shell of the Castaliini tribe is solid, subquadrangular, with an elevated umbo and a deep umbonal cavity. The umbo is variably developed, sometimes quite obsolete. Inhalant opening becomes perfectly closed at the front, exhibiting the tendency to form a true siphon. Glochidia subtriangular (equilateral or isosceles), with triangular, straight and short teeth, wide at their base but not divided at the end, cirrus grouped in the form of tufts; no larval filament. Genus Castalia, Castaliella and Callonaia.

The Diplodontini tribe is characterized by a regular-shaped shell, not winged, and always with radial ribs in the umbonal region, but with variable growth. A weakly developed posterior sulcus, with the exception of some forms that are more elongated and more acute posteriorly (e.g. *D. parallelipipedon* or *D. parodizi*). Inhalant opening similar to Prisodontini.

Their glochidia are subtriangular (scalene), with S-shaped teeth, curved and ending in a pair of spicules; they have a long larval filament; and 2-4 sensitive cirri. Species with direct development do not have teeth or hooks on the embryonic shell, but one or more marked growth bands. Genera Diplodon and Rhipidodonta.

In the tribe Diplodontini, the species show extraordinary individual and ecological variation (Parodiz, 1982), often repeated or mixed among the numerous local populations, but without taxonomic value (Ortmann, 1921; Mansur, 1972; Parodiz and Bonetto, 1963). Subgeneric divisions have been proposed based solely on transient shell characteristics. The most convincing separation is based primarily on the parasitic or non-parasitic condition of the larva; and in a second stage, groups of species can be recognized by conquiliological characters. Despite this, they sometimes present serious identification difficulties due to the intermediate variations that the groups present.

The biology, ecology, reproductive and systematic aspects of freshwater bivalves of the superfamily Etherioidea, especially in the state of Sao Paulo, are still scarce and in many respects unknown. Thus, we can mention: comparative anatomy of *Anodontites trapesialis* and *A. trapezeus* (Hebling, 1976), functional anatomy of *D. rotundus gratus* (Hebling and Penteado, 1974), *Fossula fossiculifera* (Avelar, 1993), *Castalia undosa undosa* (Avelar and Santos, 1991), *D. rhombeus fontainianus* (Avelar and Cunha, 2009), gametogenic cycle of *D. rotundus gratus* (Avelar and Mendonça, 1998).

As for the larval forms, several authors, including Ortmann (1921), Bonetto (1959a, 1959b, 1960a, 1960b, 1961a, 1961b), Bonetto *et al.* (1962), Mansur and Campos-Velho (1990), Mansur and Silva (1999) and Mansur and Campos-Velho (2000), studied some aspects of the biology, anatomy and/or meristics of the glochidia of South American Hyriidae species.

According to Mansur (1970), systematic studies have been carried out using superficial shell data, resulting in a large number of synonyms and citations of species for the same site, when in fact they are ecological variations. For Parodiz (1982), these variations do not exist, but if two or more forms of a single species are considered to inhabit the same ecological niche, they can hardly be called ecological. A known taxonomic view of the species in the area would make it possible to recognize hybrid forms.

The genus Diplodon is considered by Bonetto and Ezcurra (1963) to be the most primitive, and there is still no satisfactory classification to group the species. The great variability in the shape of the shell and the different types of environments they occupy in the vast expanse of the hydrographic network have led to discussions regarding the number of species and synonyms (Bonetto, 1951, 1961a, 1961b, 1961c; Mansur and Anflor, 1981). Studies of the internal morphology of the stomach comparing type specimens with specimens identified on the basis of conquiliological characteristics alone show that the work of determining species of the genus Diplodon can be mistaken (Mansur and Anflor, 1981).

According to Bonetto (1961a, 1961b), as well as knowing the conquiliology and morphological characteristics of the adult, the ideal way to study the systematics of the genus Diplodon is to know the morphological characteristics of the gloquidia, the parasitic stage of species of the genus, which need an intermediate host to complete the reproductive cycle. According to the author, the gloquidial characteristics are always preserved whatever the geographical distribution and environment in which the species live.

Glochidial characteristics help in the identification and classification of this group of molluscs in which juvenile or adult individuals often do not provide interspecific diagnostic criteria (Mansur and Silva, 1999).

Another approach would be through molecular analysis (Graf and O'Foighil, 2000; Graf, 2002; Graf and Cummings, 2006).

For the state of Sao Paulo, there is still no contribution to the in-depth study of the larval forms of native species. The study of the glochidia, the position and structure of the shell are easily distinguishable characters for many species (Bonetto, 1965).

Due to the numerous taxonomic controversies involving the species of the genus Diplodon, whose identification is largely related to the simple characteristics of the shell, it is becoming increasingly necessary to add knowledge of the morphology and biology of freshwater bivalves, thus justifying the study for a correct specific diagnosis and for the preservation of the species of this group, which is so interesting and at the same time so threatened.

Life cycle

The way in which organisms reproduce and disperse can have profound effects on the genetic structure and demographic process of populations (Kat, 1984). Likewise, the way in which they reproduce and disperse can play an important role in species longevity, geographical distribution and speciation rates (Mayr, 1971; Kat, 1984; Graf, 1997; Wachtler *et al.*, 2001). Limnetic bivalves of the superfamily Unionacea are the only members of the class Bivalvia that are known to have a parasitic larval stage in their life cycle, which involves an obligatory relationship with a vertebrate host, usually a fish, and a highly modified larva, the gloquid or lasid. This parasitism is considered atypical, as these organisms only have a parasitic larval stage, while the adults are free-living (Kat, 1984; Wachtler *et al.*, 2001).

During the spawning period, mature males release large quantities of gametes into the water. For fertilization to occur, these gametes must pass through the inhalant opening of sexually mature females of the same species. The male gametes pass through the opening into the suprabranchial chamber, where the eggs are kept. The fertilized eggs are transferred to the gill chambers. This chamber forms a modified incubation sac, called a marsupium. In the formation of the marsupium, both or only one pair of gills may be involved. During their stay in the marsupium, the fertilized eggs develop into embryos and metamorphose into a larval form (Mackie, 1984).

The parasitic larvae can have two general forms. While Unionoidea has only gloquidia, Etherioidea *sensu* Ortmann has both gloquidia and lasidia. The glochidia are small (70-350 μm), made up of a simple adductor muscle and mantle cells enclosed by two calcareous valves connected by a hinge (Arey, 1932). They adhere to the host tissue by closing their valves over the exposed epithelium of the gills or fins. The host tissue encysts the mollusc larva, and it is inside this cyst that the gloquid metamorphoses into a juvenile (Kat, 1984; Graf, 1997). Gloquidia generally belong to one of two varieties: (1) subcircular to suboval and without hooks or (2) subtriangular and with hooks (Coker *et al.*, 1921; Hoggarth, 1999), although there is variation within these types (*Potamilus* ; Roe and Lydeard, 1998). Glochidia can also have a structure similar to a long bisso's thread (Mansur, 1999; Wachtler *et al.*, 2001). Lasidians are also small (85-150 μm, not including the larval bisso), but they are trilobed larvae with a univalve, non-calcified shell. As with gloquids, they can take two forms: (1) lasid-like and (2) haustorian-like (Bonetto and Ezcurra, 1963). Although different in their morphology and size, the fundamental distinction between the two varieties of lasidium is that while the former adheres to the host by forming cysts, the haustorium type adheres by tubular appendages (Fryer, 1954, 1961; Parodiz and Bonetto, 1963; Wachtler *et al.*, 2001).

The gloquidia can be released after a few weeks, or the mature gloquidia are kept in the marsupium for months, depending on the species (Kat, 1984; Mackie, 1984).

The release of gloquidia can occur in different ways, reflecting the varying degrees of adaptation involved in both attracting the host fish and increasing the likelihood of adherence (Kat, 1984). In most species, the gloquid passes from the marsupium into the suprabranchial canal, and is released through the exhalant opening. When released in this way, the glochidia are more or less grouped together by a layer of mucus, which either dissolves soon after release or holds the glochidia together in so-called conglutinates, which can be of various shapes and colors (Ortmann, 1910; Arey, 1932; Graf, 1997; Pillow *et al.*, 2008). Among some species of Lampsilinae, the glochidia are released in irregular masses through special openings at the base of the marsupium, and are released through the inhalant opening by a rapid adduction of the valves (Ortmann, 1910; Graf, 2000).

After their release, the gloquids must come into contact with their hosts, the fish. After a period of incubation, which varies according to the species, the gloquids undergo metamorphosis into sessile juveniles, similar to the way adult bivalves live.

Enlistment

Unlike marine bivalves, naiads generally need a host fish in order to complete their life cycle. As parasites, a portion of their nutrition is derived from the host (Fisher and Dimock, 2002), but most unionids do not grow during encystment (Lefevre and Curtis, 1912). The attachment of gloquids does not cause any damage to the host fish (Henley and Neves, 2001), however, if parasitism occurs in large quantities, it can lead to the host's death (Silva-Souza and Eiras, 2002).

If the glochidium attaches itself to a compatible host fish, the epithelial tissue of the fish's host cells migrates over the larva, encapsulating it in a cyst, inside which the larva develops and metamorphoses into a juvenile. The larva remains encysted for days to months, depending on the species and temperature, and transforms into juvenile individuals, developing a true heart, digestive tract, and a muscular foot. When metamorphosis is complete, the juvenile detaches and begins an independent sessile life, becoming benthic suspensivores (Arey, 1932; Young and Williams, 1984a; Mansur, 1999).

Gloquids that attach to incompatible (non-host) species are detached within a few days of attachment because they either fail to encyst or are subsequently detached from the host before the transformation is complete (Rogers-Lowery *et al.*, 2007).

Kirk and Layzer (1997) and Henley and Neves (2001) showed that there is an immunological response by the fish to the encystment of the gloquid. And that when it doesn't adhere to the

specific host species, does not complete its development satisfactorily. This same species-specific response between bivalve larvae and their host fish has also been found by other authors, such as Zale and Neves (1982), Young and Williams (1984b), Kat (1984), Williams *et al.* (1993) and Wachtler *et al.* (2001).

Host specificity, from an immunological point of view, involves the adaptation of gloquids to survive the innate defensive responses of the host fish. Innate immune responses are those that do not require prior exposure of the host individual to parasite antigens. Fish can also acquire immunity to gloquids via an adaptive immune response, including the production of antibodies. However, antibody production and adaptive immunity develop slowly in fish and apparently affect gloquids usually after multiple infestations (Meyers *et al.*, 1980; Bauer and Vogel, 1987; Rogers and Dimock, 2003; Dodd *et al.* 2005, 2006).

An innate response of the host is the encapsulation of parasites fixed by epithelial cells, called keratocytes (Arey, 1932; RogersLowery and Dimock, 2006). This process is essential for successful parasitism by gloquids, but paradoxically, encapsulation seems to be an anti-ectoparasite and anti-wound response. After encapsulation, cellular defenses such as granulocytes and phagocytes are concentrated in the capsule and can kill incompatible gloquids. Incompatible glochidia are also released when the capsule degenerates or disintegrates as a small mass of tissue (Arey, 1932; Meyers *et al.*, 1980; Waller and Mitchell, 1989).

Naiads exhibit varying degrees of host specificity. The proportion of gloquids that successfully undergo metamorphosis can vary greatly between host species. Most unioniform species are not able to utilize a wide variety of fish species. In general, a few well-related host species, or populations within species, will support the development of the glochidia of several species (Zale and Neves, 1982; Neves *et al.*, 1985; Yeager and Neves, 1986).

Several studies have been carried out to identify the specific host species of these bivalves (Hoggarth, 1992; Watters and O'Dee, 1996; Haag and Warren, 2003). In contact with good hosts, 90% of fixed gloquids can successfully metamorphose into juveniles, while only a small proportion will succeed in marginal host species (Bigham, 2002; Barnhart *et al.*, 2008).

During their natural development cycle, the mortality rate of the larvae is very high due to the need to find a host fish, predation, removal to inappropriate habitat by currents, bacterial and fungal diseases during parasitism in the fish and finally settlement in sediments not suitable for the juvenile bivalve. Young and Williams (1984a), in Scotland, estimated the mortality of gloquid larvae in the wild of the perliferid bivalve *Margaritifera margaritifera* at over 99% between incubation in the female and encystment in the fish, and between detachment from the fish and settlement on the river bottom.

This characteristic of the life cycle is a major component of any conservation plan for limnetic bivalves, but the host fish for several species are not known or have not been fully investigated (Zale and Neves, 1982; Watters, 1996; Watters and O'Dee, 1998).

Conservation

The fauna of limnetic bivalves has shown a marked decline over the last century (Watters, 1996; Cosgrove and Hastie, 2001; Villella *et al.*, 2004; Christian *et al.*, 2004; Bogan, 2008), and the same has been happening with the native fauna of Brazil (Beasley *et al.*, 2000; Beasley, 2001; Almeida, 2006). The consensus is that the most dramatic cause of the decline and extinction of dulcicolous bivalves is the modification and destruction of their habitat. This can occur due to the effects of dams, channelization, changes in water depth (river dredging), changes in water flow or changes in the deposition of fine particles (silt and sand). These changes affect not only the bivalves, but also the fish that are used during the parasitic phase of the limnetic bivalves. Additional impacts include the use of water for industry and irrigation, pollution, the creation of sealed areas without percolation due to urbanization and road construction (Bogan, 2008) and global climate change (Hastie *et al.*, 2003).

The introduction of exotic (non-native) aquatic animals also poses a great danger to native freshwater molluscs. Introduced species reproduce rapidly, competing with native molluscs for food and space. A well-known example is the invasion of North American rivers and lakes by the zebra mussel, *Dreissena polymorpha* (Ricciardi *et al.*, 1998), and the recent introduction in Brazil of the species *Limnoperna fortunei* (Mansur *et al.*, 2003) and *Corbicula fluminea* (Beasley *et al.*, 2003).

An alternative would be to use the cultivation of these molluscs to maintain natural stocks. Aquaculture can help preserve biodiversity when, as a successful economic activity, it can provide local assistance to pressure on commonly exploited aquatic species (Secretariat of the Convention on Biological Diversity, 2004). For aquaculture to be sustainable, it must meet certain basic requirements, such as always producing non-negative changes in natural resource stocks and environmental quality (Sachs, 1993; Assad and Bursztyn, 2000).

To this end, some researchers in Europe and North America have started a series of studies to reproduce these freshwater bivalves and keep them in the laboratory for later release into the environment (Young and Williams, 1984; Buddensiek, 1995; Gatenby *et al.*, 1997; Kirk and Layzer, 1997; O'Beirn *et al.*, 1998; Tankersley and Butz, 1998; Neves, 1999; Beck, 2001; Henley, 2002; Zimmerman and Neves, 2002; Mummert, 2006). However, such work requires knowledge of the host fish in order to obtain the juvenile forms of the bivalves.

One way of getting around this problem would be to artificially culture the larvae of limnetic bivalves (Lefevre and Curtis, 1912; Isom and Hudson, 1982; Hudson and Isom, 1984; Young and Williams,

1984b; Keller and Zam, 1990; Hudson and Shelbourne, 1990; Uthaiwan *et al.*, 2001, 2002, 2003). The artificial cultivation of freshwater bivalves has been largely ignored. The few studies that have been carried out have been unsuccessful or prohibitive from the point of view of methodology and time consumption (Isom and Hudson, 1982; Keller and Zam, 1990). The degree of success achieved today indicates that the production of sufficient quantities of viable individuals for preservation and potential commercial exploitation is possible (Uthaiwan *et al.*, 2001, 2002, 2003).

Although research into artificial media began with mixing components based on the concentration of components found in fish plasma as a guide, Isom and Hudson (1982) also reported success using commercially available cell culture media (Eagles and M 199 media), which contain practically all the amino acids in concentrations as high or even higher than those found in fish plasma.

Heard and Hendrix (1964) found that the closure of the valve in three species of Lampsilinae was affected by blood and amino acids from various species of fish. Young and Williams (1984b) observed that the opening and closing movement of the gill valves of *M. margaritifera* increased considerably when they were brought into contact with the mucus, blood, gills or fin tissue of their host, the brown trout, *Salmo trutta*. The original medium was made from a Ringer's solution for Unionids with the essential and non-essential amino acids of Eagles containing NaHCO3 for pH control, vitamins, antibiotics, and glucose as the artificial portion, and fish plasma as the natural protein source, in a final ratio of two-thirds artificial medium to one-third plasma (Isom and Hudson, 1982). Other formulations were tested by Hudson and Shelbourne (1990) in an effort to improve the results in more species of bivalves, resulting in a comparison of 64 combinations of different media. Keller and Zam (1990) reported the transformation of gloquids into juveniles *in an in vitro* culture medium that included only serum. Hudson and Shelbourne (1990) showed that the transformation rate for gloquidia increased with the inclusion of fish plasma. Hudson and Isom (1984) reported the importance of adding fish plasma to an artificial medium, as well as demonstrating that the transformation of gloquidia could occur in an artificial medium with the plasma of unconfirmed host fish species. Uthaiwan *et al.* (2001, 2002, 2003) developed a simple technique for culturing gloquidia in artificial media, which resulted in a high metamorphosis rate, working with *Hyriopsis myersiana*. Similar results were obtained with the European species *Anodonta cygnea*, which showed a larval survival rate of 34.3%, while the proportion of larvae that underwent metamorphosis was 60.8% (Lima *et al.*, 2006).

Tankersley's work (2000) indicates that the levels of lipids in the gloquids and juveniles vary according to each broodstock *pool*, and are therefore influenced by the parental nutritional status.

In addition, he indicates that the type of culture medium used *in vitro* influences the lipid content of the gloquids and subsequent juveniles. In *Utterbackia imbecillis*,

Fisher and Dimock (2006) found that larvae adhered to the host fish accumulated lipids and glycogen in the larval mantle cells, while larvae reared *in vitro* did not show this accumulation.

Recovery projects for endangered bivalve species recommend propagation and reintroduction as a way of conserving populations (Mummert, 2006). Due to the complexity of the life cycle of nannies, artificial cultivation (*in vitro*) could offer ways to overcome the obstacles present in the natural life cycle of these animals, thus contributing to the maintenance of biodiversity in limnic environments.

Research objectives

In Brazil, the indigenous bivalve fauna is currently suffering from anthropogenic impacts and interspecific competition due to the invasion of the exotic species *Corbicula fluminea* and *Limnoperna fortunei*. These bivalves, of Asian origin, compete for both habitat and food. In addition, their reproductive mechanisms, which do not involve a host species and are continuous throughout the year, favor the invasive species over the native ones.

In order to propagate the species that are in danger of extinction, we intend to develop viable techniques for obtaining juvenile forms in the laboratory, through the artificial cultivation (*in vitro*) of gloquidia, the name given to the larval forms of the genus Diplodon. At the end of this stage, we will be providing the scientific community with parameters for future research aimed at preservationist actions, in an attempt to reconstitute the bivalve fauna characteristic of each of the river basins.

In order to achieve the objective of carrying out *in vitro* reproduction of Diplodon genus naiads, the research was divided into phases.

1. Research phases

Figure 3 illustrates the research stages. The first phase, defining the collection site, consisted of several preliminary collections in rivers near Ribeirao Preto, with the aim of identifying a bivalve population that was easy to access and of a good size. Once the study site had been defined, a bibliographical survey of the region was carried out, as well as citations of the occurrence of species of the genus Diplodon.

From this point onwards, during the second phase, the research was twofold: one prioritizing the identification of adult specimens, and the other focused on propagating the species in the laboratory.

To identify the specimens collected, we searched for papers with the original descriptions of the species, thus creating a database with the literature consulted, but not always cited.

Due to the lack of information on the study group, both bioecological and for the management of these bivalves in laboratories in South America, it was necessary to develop a protocol with new methodologies to achieve success in the reproduction of this group under artificial conditions. With this information in mind, the methodology was systematized chronologically. Not all the variables tested could be applied simultaneously. In order to obtain a protocol, the treatment that resulted in the greatest success and provided the best survival of the semaphores was always maintained. It was therefore not possible to check the interactions between the variables simultaneously.

In the final stage, the results were generated, represented by graphs, tables and images, as well as their interpretation, with the aim of adding a little more knowledge about the fauna of limnetic bivalves of the Hyriidae family in the state of Sao Paulo.

Figure 3: Research phases and work stages.

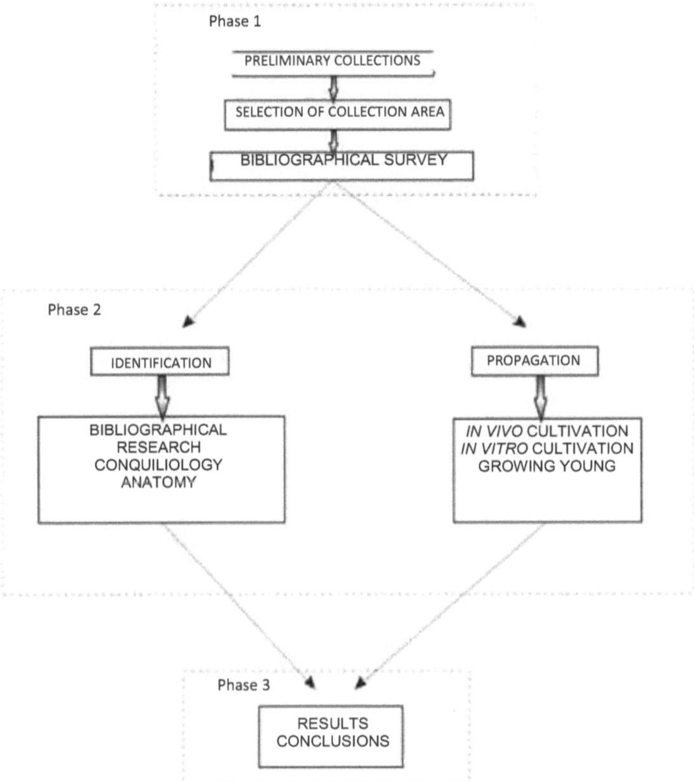

2. Collection

In order to locate an established bivalve community, a series of preliminary collections were made in the Mogi Guaçu, Pardo and Sapucaí rivers, in a reservoir and in a lake. The coordinates of the collection points are listed in Table 1.

Table 1. Locations and coordinates where the bivalves were collected.

LOCALITY	MUNICIPALITY	COORDINATES
		21°55'14.4" S and 47°23'05.1" W
	Pirassununga	21°55'39.1" S and 47°22'44.0" W
		21≡55'41.3" S and 47222'31.3" W

Mogi Guaçu River		21255'41.3" S and 47°22'22.1" W
	Luiz Antonio	21°36'45.2" S and 47°49'09.6" W
		21°50'36.1" S and 47°29'44.5" W
	Porto **Ferreira***	21°49'62.1" S and 47°30'06.1" W
Beija Flor Stream	Luiz Antonio	21°36'54.1" S and 47°48'52.4" W 21°36'50.1" S and 47249'01.2" W
Rio Pardo	Serrana	21°10'44.9" S and 47°34'30.5" W
	Jardinópolis	21°04'02.4" S and 47°49'27.2" W
	Ribeirao Preto	21°07' S and 47°45' W
Sapucai River	Sao Joaquim da Barra	20231'34.8" S and 47°49'42.9" W
Galo Bravo Reservoir		21°07'06.9" S and 47°49'32.1" W
Lake Monte Alegre (USP)	Ribeirao Preto	21211' S and 47°43' W

* coordinates where the study was carried out

The collections were carried out with permission granted by the Brazilian Institute for the Environment and Renewable Natural Resources - IBAMA (# 10207/2007).

Of the points sampled, the Mogi Guaçu River, in the municipality of Porto Ferreira, state of Sao Paulo, Brazil, was chosen as the study site.

At the collection site, abiotic factors were measured: water conductivity using a YSI Model 30 Handheld Salinity, Conductivity and Temperature System; dissolved oxygen and temperature using a YSI Model 52 D.O. Meter; and turbidity using a Model 2100P Handheld Turbidimeter. Data on the flow of the Mogi Guaçu River at the collection point was taken from the National Water Agency (ANA) database.

The bivalves were collected by groping the bottom of the river with their hands and feet, and when necessary with the help of diving equipment, since they live buried in the sediment and there is a great variation in the river's water column, especially during periods of heavy rainfall.

The specimens collected were individually packed in 200 ml wide-mouthed bottles, placed in a cooler with ice and transported to the Malacology Laboratory of the Biology Department at FFCLRP-USP (figure 4), where they were kept alive, individually, in aquariums at a constant temperature. At first, due to the lack of suitable equipment for acclimatizing the animals, the temperature varied between 22 °C and 26 °C, but after improving the air conditioning system it was possible to maintain a temperature of 20 °C.

Figure 4 - Methodology for transporting the collected specimens.

3. Identification

All the material collected was sorted in the laboratory; some specimens were fixed in 70% alcohol (shell and soft parts) while for others, only the shell was preserved. All the animals were measured and photographed and, where possible, identified to species level using the works of Ortmann (1921), Mansur et al. (1987) and Simone (2006) as a reference.

The collected material was first identified by separating the bivalves by shell morphology and shape, external sculpture and umbonal sculpture (figure 5).

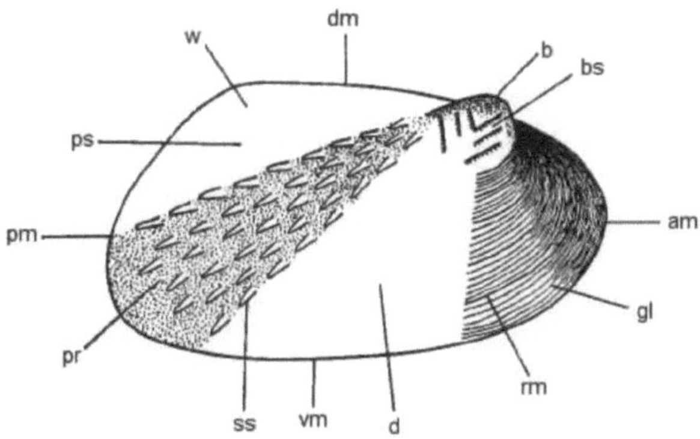

Figure 5 - Diagram of the external characteristics of the shell of a hypothetical freshwater bivalve. Legend: am, anterior margin; b, beak (umbo); bs, umbonal sculpture; d, disc; dm, dorsal margin; gl, growth lines; pm, posterior margin; pr, posterior rib; ***ps****, posterior depression;* ***rm****, growth lines;* ***ss****, shell sculpture;* ***vm****, ventral margin;* ***w****, wing. Modified from McMichael and Hiscock (1958).*

In a second stage, some specimens were dissected to observe some macroscopic anatomical features, such as the inhalant and exhalant opening, gills and palps.

The identification of the specimens, collected in the Mogi Guaçu river, in the municipality of Porto Ferreira, was confirmed through visits to the malacological collection of the Zoology Museum (MZUSP), under the supervision of Prof. Dr. Luiz Ricardo Lopes Simone, and consultation of the collection of the BIOTA/FAPESP project in the Malacology laboratory of the Ribeirao Preto Faculty of Philosophy, Sciences and Letters.

The duly identified specimens will be deposited in the Zoology Museum (MZUSP) and in the collection of the Ribeirao Preto School of Philosophy, Sciences and Letters.

4. Bivalve propagation in the laboratory

4.1. Phytoplankton production

4.1.1. Isolation of microalgae strains

Samples of phytoplankton were collected from the Mogi-Guaçu River (21°50'36.10" S and 47°29'44.5" W), the Galo Bravo Reservoir (21°07'06.9" S, 47°49'32.1" W) and Monte Alegre Lake, on the Ribeirao Preto/USP Campus, using a 30 µm mesh plankton net. The samples were placed in a culture medium in order to identify the phytoplankton composition and thus select possible species that could be cultivated and were part of the bivalves' diet. Through repeated dilutions and with the aid of a micropipette, the following genera were selected for cultivation: *Chlamydomonas* sp, *Chaetoceros* sp, *Monoraphium* sp, *Chlorella* sp, *Nitzschia* sp, *Phaeodactylum* sp and *Scenedesmus* sp.

The strains of the microalgae species were replanted every 4-5 days, always respecting the initial concentration of 1/5 inoculum for the final volume (e.g. 1 liter of inoculum/4 liters of culture medium). The culture medium used to maintain the strains was autoclaved (20 minutes at 120 °C) before being used, while the mass production culture medium (volume > 5 liters) was left untreated.

4.1.2 Culture medium for growing microalgae

Two fertilizers were tested as a growing medium: cattle manure and an inorganic fertilizer (NPK) (Sempre Verde® agricultural fertilizer, donated by Bonigo Ltda), with nitrogen: phosphorus: potassium concentrations of 12:6:6.

The organic growing medium was prepared using dried, tanned manure. It was sieved through a 125 µm mesh and placed in an oven at 80 °C for 24 hours. After this period, the manure was diluted in distilled water (45 grams of manure/5 liters of water) and subjected to constant aeration for 24 hours. The growing medium was then filtered through a 100 µm mesh and used to grow the microalgae.

The growing medium enriched with inorganic fertilizer was prepared in different dilutions with distilled water: 0.5/200; 1/200; 5/200; 10/200 and 20/200 (Table 2).

Table 2. Concentrations (g/ml) of phosphorus, potassium and nitrogen in each dilution tested for phytoplankton cultivation.

	Concentration (g/ml)		
Proportion	**N**	**P**	**K**
0,5/200	0,18 x 10^{-4}	0,09 x 10^{-4}	0,09 x 10^{-4}
1/200	0,36 x 10^{-4}	0,18 x 10^{-4}	0,18 x 10^{-4}
5/200	1,8 x 10^{-4}	0,9 x 10^{-4}	0,9 x 10^{-4}
10/200	3,6 x 10^{-4}	1,8 x 10^{-4}	1,8 x 10^{-4}
20/200	7,2 x 10^{-4}	3,6 x 10^{-4}	3,6 x 10^{-4}

Legend: **N**, nitrogen; **P**, phosphorus; **K**, potassium.

After preparing the growing medium, 20 ml of the microalgae strain was added to a 250 ml flask. The cultures were exposed to uninterrupted fluorescent light (04 white light bulbs 36 W), with constant aeration.

The cultures were analyzed daily for eight days. A 5 ml aliquot was taken from each culture and a drop of lugol (4%) was added to fix the microalgae. Eight readings were made for each aliquot, using a hematocimeter with a Neubauer chamber. The daily cell density result was determined by the arithmetic mean of the eight cell counts, obtained using the formula:

DC = cel x $_{104}$ x Dilution Factor,

Where **DC** corresponds to cell density (cells/milliliter) and **cel** corresponds to the number of cells counted.

The cell density values for each treatment were compared over the eight-day experimental period by an Analysis of Variance (ANOVA), followed by a Tukey test, at a significance level of 95%. The mathematical transformations, where necessary, and the ANOVA were carried out using the Statistica program (StatSoft, Inc., 1998).

4.1.3. Microalgae maintenance

The microalgae were grown in 10-liter flasks, kept in an air-conditioned room at a temperature of 20 °C, with constant aeration and light intensity (04 white light fluorescent lamps 36 W), without photoperiod, in non-axenic conditions (figure 6). The algae strains were repotted every 7 days and inorganic fertilizer was used, with nitrogen: phosphorus: potassium concentrations of 12:6:6, in a ratio of 1/200.

Figure 6: Maintenance of the microalgae strains, with constant light and aeration.

We kept growing phytoplankton colonies of the species *Chlamydomonas* sp, originally from the Mogi Guaçu river and isolated in the laboratory; and *Ankistrodesmus* sp, kindly provided by Prof. Dr. Sonia Helena Sipauba, from CAUNESP/UNESP, Jaboticabal, and cultivated according to the methodology already described.

4.2. Keeping adult bivalves in the laboratory

4.2.1. Preliminary experiments

To develop a methodology for keeping adult bivalves in the laboratory, we used the species *A. trapesialis*, which inhabits lakes and ponds in abundance, thus not increasing the pressure that already exists on the endangered species of the genus Diplodon.

Specimens of *A. trapesialis* were collected in the marginal region of the Galo Bravo lagoon (21°07'06.9" S and 47°49'32.1" W), in the municipality of Ribeirao Preto, SP. The specimens were placed in a thermal box, with enough local water to cover them, and transported to the Malacology laboratory of the Biology Department at FFCLRP-USP.

The bivalves were subjected to two treatments. One group (N=50) was placed in 32x28 cm nylon bags (10x3 mm mesh opening), with three animals each. The bags were kept suspended in a 250 l fiber cement box with mine water and constant aeration. Another group (N=50) was placed in another 250-liter fiber cement box.

I, with a 10 cm layer of silt (fine sand) (figure 7).

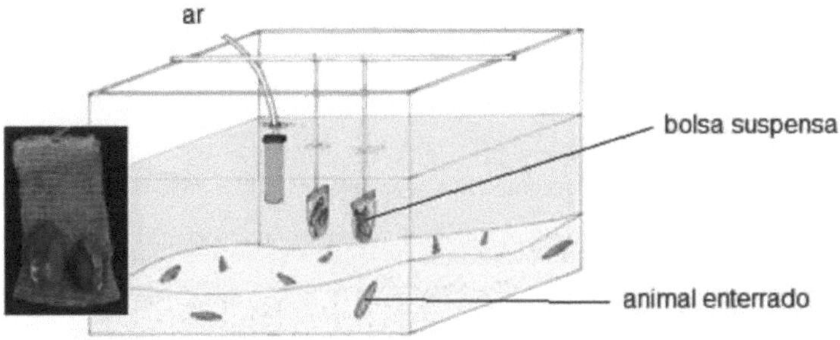

Figure 7. Diagram of the maintenance unit where the specimens of Anodontites trapesialis were kept suspended and buried. Photo detailing the nylon bag with two individuals.

A diet of microalgae (*Chlamydomonas* sp) was offered daily *ad libitum*, and the water in the boxes was changed weekly. Each month the animals were removed from the water and weighed on a Gehaka BG 400 electronic scale, accurate to 0.01g.

In the statistical analysis, the mortality of the two groups was compared using Pearson's chi-square test. To evaluate growth in weight, a repeated measures ANOVA model was used considering two factors: group with 2 levels (buried and suspended) and time of evaluation with 5 levels (initial, 1st, 2nd, 3rd and 4th month). Comparisons between groups and within groups were evaluated using orthogonal contrasts. The significance level adopted was 0.05. The statistical program used was SAS, version 9.0.

4.2.2. Maintenance of collected Diplodon bivalves

The specimens transported to the laboratory that had not released gloquidia after five days were suspended in 32x28 cm nylon bags (10x3 mm mesh opening), with two animals each, at 15 cm intervals between the bags, 50 cm from the bottom of the maintenance unit. The unit consists of a 250-liter fiber cement box, with a continuous water flow system (average flow of 10.0 L/min), located in an environment with a constant temperature (20 °C) and a 12-hour light and 12-hour dark cycle. Two 5.0 cm air diffusers were used to suspend the phytoplankton algae in the water column.

The bivalves were fed the unicellular green algae *Chlamydomonas* sp and *Ankistrodesmus* sp (in a 1:1 ratio), at a density of 400,000 cells/ml, once a day, and the water was changed and the animals cleaned monthly. The choice to use these species of microalgae was due to their easy handling and rapid growth in our laboratory. In addition, both have an appropriate size for larvae and adults.

The water used for phytoplankton cultivation and for the maintenance unit comes from a mine system on the campus of the University of São Paulo in Ribeirao Preto.

4.3. *Obtaining gloquidia*

4.3.1. Release

The animals, recently collected and previously acclimatized, were placed, still individually, in a 50 l aquarium, with mine water and constant aeration, at an ambient temperature of 20 °C (figure 8). When the larvae were eliminated, the animals were separated and the eliminated gloquidia, wrapped in mucus, were removed using a pipette and placed in a 100 ml beaker for later use. The animals that released the larvae were preserved in 70% alcohol for correct identification and those that were not frozen were kept in the laboratory and later returned to the collection site.

Figure 8. Conditioning of the newly collected animals, individually, for observation of elimination and collection of gloquidia.

4.3.2. Viability of gloquidia

The viability of the gloquidia was tested using a solution of sodium chloride on a sub-sample of gloquidia from each treatment (Lefevre and Curtis, 1912). When viable, the gloocysts are positioned with their valves open, but when exposed to a saline solution, they contract the adductor muscle, closing their valves. From the sub-sample, an initial count of 25 glochidia was made. After the addition of the saline solution, and after 10-20 seconds, the 25 glochidia were recounted and the number of open and closed individuals recorded. The animals that remained open after the addition of sodium chloride were considered to be functionally dead.

The survival of the gloquids (live/dead) was modeled by a logistic regression analysis, and no correlation was necessary as new gloquids were used in each sampling (Zimmerman and Neves, 2002). Three replicates, in 100 ml beakers, were used for each temperature and for each gloquid treatment.

The gloquidia from each treatment were subjected to different temperatures: in an incubator at 18 °C, in a temperature-controlled room at 22 °C and at room temperature 25 ± 1 °C.

The exposure modes included a treatment with mucus, in which the gloquidia were used directly after natural release; and a treatment without mucus, which refers to those gloquidia that were expelled and then processed to remove the mucus.

One of the processes used was the boiling of the newly collected gloquids, through careful and repeated changes of the water in the beaker. Autoclaved mine water, distilled water and deionized water were used. Another process was to carry out this washing on a nylon net with a mesh opening of 50 µm. As a final treatment, the gloquids were washed with an antibiotic and fungicide solution from Invitrogen® (Product code 15240-062).

4.3.3. Gloquid conchylometry

To measure the glochidia, after separating the mucus by repeated washing with distilled water, permanent slides were prepared according to the methodology of Mansur and Campos-Velho (1990), which consists of removing the glochidia tissues with eight drops of commercial sodium hypochlorite 5% NaOC in 10 ml of distilled water for five minutes, after which the material was washed with successive changes of distilled water.

After rapid dehydration in a series of 50 %, 60 %, 70 %, 80 %, 90 % and 96 % ethyl alcohol, the material was subjected to two baths in isopropyl alcohol, and finally the valves were pipetted onto laminates and mounted with Entelan and covered with laminates.

The gloquidia were observed and photographed. The image capture procedure was carried out at the Laboratory of Neurobiology and Venoms, run by Prof. Dr. Wagner Ferreira dos Santos, using a system consisting of a Leica digital color camera (DFC300 FX) connected to a microscope (DM 5000 B, Leica Microsytems - Germany) and a computer. The Q-Win program (Leica Microsytems - Germany) was used to measure the glochids. Measurements were made according to Bonetto (1959b), where transects are drawn over the image of the glochidia in order to measure the greatest length, height, length of the dorsal line, displacement of the ventral tip in relation to the center of the dorsal line, and the angle of obliquity. The latter is the measurement between the line joining the ventral tip to the middle of the dorsal line and the perpendicular from this line towards the ventral tip (figure 9). We used 40 gloquids from each spawned female.

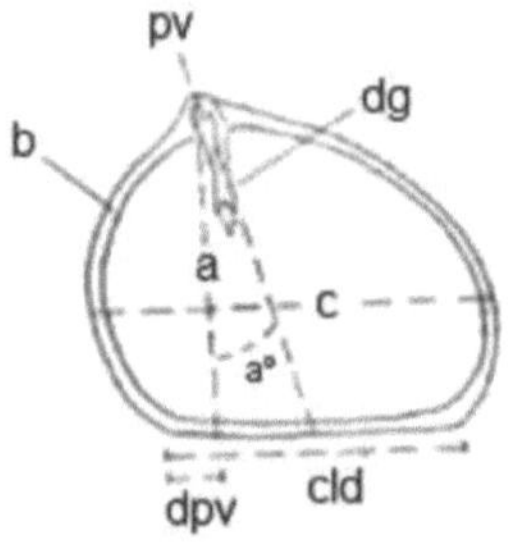

Figure 9. Conchylometry of the gloquidial valve. Legend: a, height; a°, angle; b, edge; c, length; cld, length of dorsal line; dg, gloquidial tooth; dpv, displacement of ventral tip; pv, ventral tip. Modified from Mansur and Campos-Velho (1990).

4.4 *In vivo* cultivation of gloquidia

As gloquids are parasites of fish, laboratory experiments were carried out to find hosts that could potentially be infested by the larvae of the bivalve species collected, of the genus Diplodon. Most of the fish were collected from the Mogi Guaçu river, in the municipality of Porto Ferreira, Sao Paulo, Brazil. They are: *Astyanax altparanae*, *Gymnotos* sp, *Randia queli*, *Hyphessobrycon eques*, *Cheirodon* sp, *Geophagus brasiliensis*, *Poecilia sp* and *Papiliochrommis ramirezi*. The exotic species *Oreochromis niloticus* was also used. The size of the fish ranged from 3 to 7 cm.

The naturally released glochidia were removed with a pipette and washed to remove mucus and other residues.

Artificial infestation was carried out using, where possible, three fish of each species. To do this, around 200 gloquids/fish were placed in contact with the fish in aquariums measuring 40.00 x 20.00 x 15.00 cm with 10 l of mine water, properly aerated, at a controlled room temperature of 20, 22 and 24 °C (figure 10). Every two days, the bottom of the tank was siphoned to detect the presence of juveniles, 2/3 of the tank's water was changed and the fish were fed commercial feed.

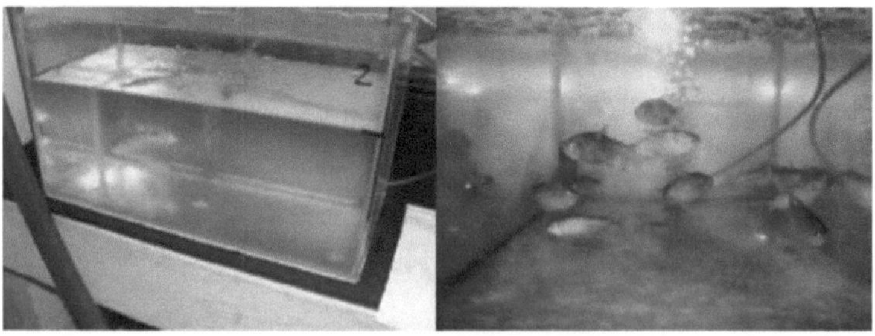

Figure 10. Artificial infestation of fish, Astyanax altparanae, with Diplodon genus gloquids (in vivo culture).

4.5. *In vitro cultivation of gloquidia*

4.5.1. *Culture medium (M 199 + antibiotic/fungicide)*

To make 1 liter of M 199 culture medium solution, a packet of M 199 powder (Invitrogen®, Cod. prod. 31100-035) was placed in a 1 liter beaker with deionized water on a shaker. Two grams of NaHCO3 were added and the pH of the culture medium was adjusted to 6.8, the value found at the point where the bivalves were collected, using 1.25 M NaOH or 1 M HCl (Keller and Zam, 1990). After the pH had been adjusted, the solution was filtered using 0.45 and 0.22 µm cellulose filters, respectively. The solution is viable for weeks, provided it is kept under refrigeration (6-8 °C) and aseptic conditions.

In the experiments during the 2008 season, a cap solution of HEPES at a concentration of 5960 mg/l was added to the culture medium before filtering, as its use eliminates the need for a CO2-enriched atmosphere (Prof. Dr. Maria Cristina da Silva Pranchevicius, personal communication). However, in 2009, after an exchange with North American researchers, HEPES was removed from the formulation, because according to Lee and Mora (2005) and Owen *et al.* (2009) this buffer solution has a toxic effect on gloquids.

A combination of antibiotic and fungicide was used to inhibit contamination of the cultures, from Invitrogen®, listed in Table 3. The solution was frozen in 2 ml aliquots, and only the amount needed was thawed at the time of use.

Table 3. Combination of antibiotic and fungicide in Invitrogen's product formulation (Product code 15240-062).

Compound	Concentration (µg/ml)
Antibiotics	
Penicillin G (IU)	10000
Streptomycin sulphate	10000
Antimycotics	
Amphotericin B	25

4.5.2. *Protein source*

4.5.2.1. *Fish plasma*

Plasma was tested from fish specimens that occur at the bivalve collection site and from exotic species. The species *Oreochromis* sp, *Colossoma* sp and *Cyprinus* sp were kindly provided by CAUNESP/UNESP in Jaboticabal, through Prof. Dr. Joao Batista Kochenborger Fernandes. The regionally occurring specimens, *Astyanax altparanae*, were purchased from a store that sells live bait.

To obtain the plasma, the fish's blood was withdrawn using a syringe with a no. 22 needle (0.70x25) from the tail vein and placed in tubes which had previously been washed with sodium heparin (2000 IU). The blood was transported in cooler boxes with ice to the Plant Molecular Biology Laboratory run by Prof. Dr. Maria Helena de Souza Goldman, where it was centrifuged at 1000 rpm for 10 minutes, then transferred to another tube and centrifuged again at 3000 rpm for 10 minutes. The plasma portion was separated and filtered through 0.45 and 0.22 µm cellulose filters, respectively, and frozen for later use.

4.5.2.2. Freeze-dried fish extract

Three specimens of *A. altparanae* (average size 5 cm) were macerated. The material was then freeze-dried at -40°C and stored for later use in a refrigerator (6-8°C). An aliquot of 0.6 mg of the freeze-dried material (0.3% of the freeze-dried fish extract after filtering) was resuspended in 1000 ml of deionized water and vortexed for 4 hours. The suspension was filtered through 0.45 and 0.22 µm cellulose filters, respectively, and stored in the refrigerator until use.

4.5.2.3. Bovine plasma

Beef blood from the Oranges slaughterhouse (Sertaozinho, SP) was placed in vials previously washed with a solution of sodium heparin. The blood was transported in cooler boxes with ice to the Plant Molecular Biology Laboratory run by Prof. Dr. Maria Helena de Souza Goldman, where it was centrifuged at 3000 rpm for 10 minutes, then transferred to another tube and centrifuged again at 4000 rpm for 10 minutes. The plasma portion was separated and filtered through 0.45 and 0.22 µm cellulose filters, respectively, and frozen for later use.

4.5.3. Growing gloquidia

The glochidia were separated from the mucus by repeated washing with distilled water and their viability was tested under light microscopy (100x) by observing the opening/closing of the valves. Around 50-100 glochidia were placed in Petri dishes (60 X 15 mm) under sterile conditions, with a culture medium consisting of M 199 solution (Sigma®), a protein source and antibiotic/antimycotic, in a ratio of 2:1:0.5.

The Petri dishes were kept in a temperature-controlled incubator (figure 11) until they were transformed into juveniles (observing the distension of the foot out of the shell is an indicator), changing the culture medium every 6 days and adding 1 ml of distilled water 24 hours before metamorphosis (Uthaiwan et al., 2002, 2003; Lima et al, 2006). Experiments were carried out at temperatures of 18, 20 and 23 °C.

Figure 11. Maintenance of gloquidia in an incubator with artificial culture medium containing M 199, a protein source and antibiotics/antimycotics (in vitro cultivation).

When possible, a completely randomized arrangement between treatments was used and eight replicates were used. The data was analyzed using an Analysis of Variance (ANOVA), followed by a Tukey test, at a significance level of 95%. The mathematical transformations, where necessary, and the ANOVA were carried out using the Statistica program (StatSoft, Inc., 1998).

4.6. Growing juveniles

After the metamorphosis of the gloquids, incubated for 18-22 days, the newly metamorphosed juveniles were removed from the culture medium, washed with a mixture of autoclaved mine water and M 199, and subjected to various treatments in Petri dishes (60x15 mm) and 150 ml Beckers, with aeration (figure 12).

Figure 12. Cultivation of juveniles of the genus Diplodon. On the left in Petri dishes, and on the right in a Becker with aeration.

The juveniles were kept in autoclaved mine water, without food supplementation; with the addition of 15,000 cells/ml of the microalga *Chlamydomonas* sp every other day; with a 1 mm layer of fine sandy sediment (particles < 125 µm); and in a final treatment, with sediment and food.

Every other day, the juveniles were observed under a stereoscopic magnifying glass, and the dead individuals were removed and counted. The survival percentage of the juveniles was transformed mathematically and the treatments were compared using a T-test.

4.6.1. Post-metamorphic juveniles

The newly metamorphosed juveniles were subjected to the treatments mentioned above immediately after metamorphosis. The number of individuals in each treatment was calculated by dividing the total number of metamorphosed juveniles (*in vivo* and *in vitro* culture) by the number of treatments. When there was little availability of metamorphosed juveniles, only one treatment (food or sediment, Petri dish or Becker) was tested at a time, with each batch.

4.6.2. Juveniles 5 days old

In subsequent releases of gloquids, the treatment with the highest previous survival rate was maintained and after five days the treatments were repeated (addition of food, presence of sediment and type of container).

4.6.3. Juveniles 15 days old

Same as above, but starting treatments after 15 days.

4.6.4. Juveniles aged 30 days or more

The 30-day-old juveniles were kept in a 150 ml beaker with 1 mm of sandy silt, with constant aeration, a natural photoperiod and a temperature of 20 °C. Every two days, 2/3 of the water in the beaker was changed and microalgae added.

The concentration of the food varied from 15,000 to 90,000 cells/ml as the juveniles grew. From the age of 60 days, the species

Ankistrodesmus sp, offered in a 1:1 ratio with the other microalgae.

RESULTS

1. Collection

1.1. Characterization of the collection site

At the federal level, the Mogi Guaçu river basin belongs to the Paranà Hydrographic Region, which is made up of the Paranà river basin, located on national territory. In the state of São Paulo, the Mogi Guaçu river basin (UGRHI- 09) is located in the northeast of the state. This UGRHI has boundaries with the UGRHI's: Piracicaba/Capivari/Jundiai; Tietê/Jacarei; Tietê/Batalha; Turvo/Grande; Baixo Pardo/Grande; and Pardo (figure 13).

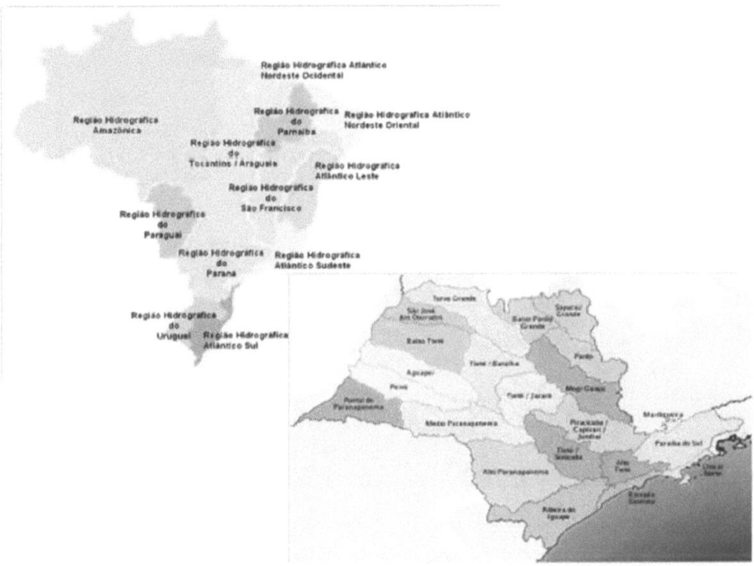

Figure 13. Location of the Mogi Guaçu River Basin. Source: http://mapas.znc.com.br/sos_bacias_sp/index.php

The Mogi Guaçu River rises in the municipality of Bom Repouso (22°28'15" S, 46 8'42" W), in the south of the state of Minas Gerais, Brazil, and its watershed has a total drainage area of 18938 km^2 (CORHI, 1999). Its source is located on Monte Curvado, 1700 meters above the sea. Its course is 473 km long, covering 95.5 km in the state of Minas Gerais, through a gorge in the Serra da Mantiqueira, at an average altitude of 825 meters above the sea. It then runs for another 377.5 km in the state of Sao Paulo, flowing into the Pardo River, 490 meters above sea level, in the municipality of Pontal (21°1'22" S, 48°2'13" W). Its main tributaries on the right bank are the rivers: Onça, Itupeva, Claro and Jaguari Mirim; on the left bank, the rivers: Eleutério, do Peixe, do Roque, Bonito, Araras and

Mogi Mirim.

The compartment with the highest drainage area density is the Peixe River, followed by the Jaguari Mirim River compartment, both located in the highlands in areas of undulating to strongly undulating relief on the Atlantic Plateau. The lowest drainage area density is precisely in the Lower Mogi compartment, a region with mostly gentle undulating terrain, intensively used for agriculture.

The chosen collection point has partly preserved banks, but most of them are well degraded with the placement of tires, PVC pipes from clandestine drains and garbage (figure 14).

Figure 14. Site where the samples were taken (Mogi Guaçu river, municipality of Porto Ferreira, SP, 21°50'36.1" S and 47°29'44.5" W). Photos A and B were taken during the dry season, and photos C and D during the rainy season.

1.2. *Anthropogenic action in the Mogiana basin:*

The Tupi-Guarani arrived in the Mogiana Basin between the 16th and 17th centuries, mainly in the region of Cachoeira de Emas, where they remained until around 1880.

The first mapping of the Mogi Guaçu river dates from 1766 and the original document is in the MZUSP in Sao Paulo. The second mapping of the Mogi dates from 1773 and the original document is in the Colonial Historical Archive, Torre do Tombo, Lisbon, Portugal.

The Pirassununga region began to be settled by white pioneers in 1809. At the beginning of the occupation, the region had a vegetation cover of 2/3 Atlantic forest and 1/3 cerrado grassland. In 1999, with around 65,000 inhabitants in the region, only 1/3 of the original vegetation cover remained.

Since the 19th century, the cultivation of coffee in the region has led to the clearing of native forests. In 1878, the railroad arrived and, with it, new forests were cleared (for firewood as locomotive fuel, sleepers, laying tracks, etc.),

From 1883 until 1917, river navigation began along a 200 km stretch between Porto Ferreira and Pontal, at the mouth of the Mogi Guaçu. Along this stretch, 15 river ports were built and, at the height of their activity, there were 11 steamers and 54 barges for transporting passengers and cargoes of goods.

At present, the Mogi Guaçu River is characterized by the absence of industrial and mining activity capable of altering the hydrosedimentological dynamics. However, its riparian forest is compromised by the irregular allocation of ranches (Neves, 2005).

1.3. *Abiotic factors*

The abiotic parameters measured, conductivity, temperature, dissolved oxygen and turbidity, show a marked variation depending on the time of year (appendix B). It is possible to visualize the flood period not only by the increase in the river's flow (figure 15).

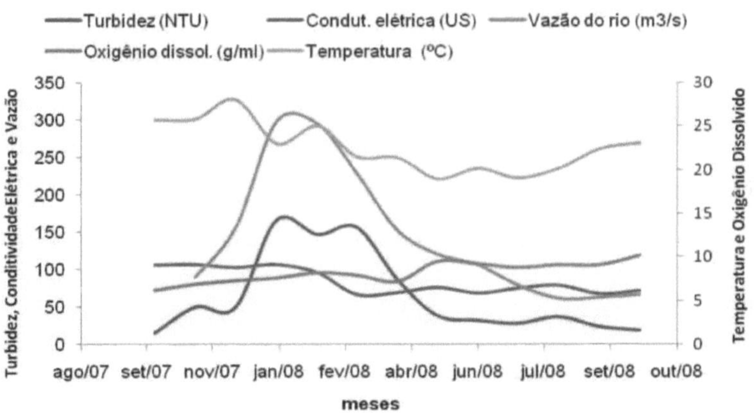

Figure 15. Graphical representation of the variation in abiotic parameters between October 2007 and October 2008. The river flow values correspond to a 10-year average (1996 to 2006).

1.4. *Occurrence of the genus Diplodon in the Mogi Guaçu River*

Ihering (1893) described the occurrence *of D. fontainianus, D. caipira* and *D. paulista*, in Piracicaba; and *D. martensi* in the state of Sao Paulo, but without specifying the hydrographic basin. He also described a new species, *D. greeffeanus*, in the type locality of Campinas and Piracicaba. Simpson

(1914) described *D. trivialis* for the Jaboticabal region. For the same locality, Ortmann (1921) described a new species, *D. mogymirim*.

Ortmann (1921) extended the occurrence of *D. paulista* and *D. fontainianus* to the river basins in the north of the state of São Paulo.

Bonetto (1959, 1961), studying Diplodon specimens from the same basin in the Pirassununga region, SP, found the same species mentioned by Ortmann and extended the distribution of *D. expansus* (Küster, 1856) from the Conigo River, Nova Friburgo (RJ), to the Pirassununga region. He also draws attention to the occurrence of *D. paulista* and *D. fontainianus*, initially described for the Piracicaba region, in Cachoeira das Emas, Pirassununga.

Parodiz (1973) reports a high abundance of *D. expansus* in the region of Piracicaba, Pirassununga, and the south of Minas Gerais. He also mentions the occurrence of *D. paulista* in the Mogi Mirim river, a tributary of the Mogi Guaçu river.

Between 2000 and 2002, the BIOTA/FAPESP project collected several species of the Hyriidae family in the Mogi Guaçu River. Among the species of the genus Diplodon, *D. fontainianus*, *D. greeffeanus*, *D. rotundus gratus* and *D. mogymirim* were reported.

2. Identification of the bivalves collected

A total of 510 live specimens belonging to two families and eight species (Table 4) were collected from the Mogi Guaçu River between March 2008 and September 2009. The species richness of each family in the central region of the Mogi Guaçu basin is represented by the Hyriidae (five species) and Mycetopodidae (two species).

Among the species collected, *D. fontainianus*, *D. expansus*, *D. rotundus gratus*, *D. martensi*, *Diplodon* sp, *C. undosa undosa*, *F. fossiculifera*, and *A. trapezeus* were identified. The conquiliological characters of each species are listed below.

Diplodon expansus (Küster, 1856) (figure 16A)

Diagnosis: Elliptical, equivalent shell, with the anterior margin usually rounded from the end of a very short lunula to the ventral margin.

Left valve with pseudocardinal teeth divided into two pieces, the anterior one larger with a crenulated edge, and the posterior one being a tooth with an acute tip; and with two parallel and arched lateral teeth, of which the lower one is wider. The right valve has its pseudocardinal teeth bifurcated in an oblique longitudinal direction. The color of the periostracum is brown, but it becomes quite dark near the margins.

Diplodon rotundus gratus (Lea, 1860) (figure 16B)

Diagnosis: Equilateral, suboval shell with straight dorsal and posterior edges. The shell has a fragile appearance, with the teeth not so developed, with one pseudocardinal tooth and two lateral teeth on the left valve and two pseudocardinal teeth and two lateral teeth on the right valve. Maximum size 4.50 cm. The periostracum is dark greenish-brown in color. The inner surface is smooth and whitish, with some purple iridescence.

Diplodon fontainianus (Orbigny, 1835) (figure 16C)

Diagnosis: Shell oval to subtrapezoidal, almost circular, with the largest diameter in the middle portion of the shell. Dorsal line strongly curved, dorsal line curved behind the umbo. Left valve with one or two pseudocardinals and two curved lateral teeth; right valve with two elongated pseudocardinal teeth, the lower one being larger, and one lateral tooth. Periostracum dark brown to black, with a bluish-white pearly inner surface.

Diplodon martensi (Ihering, 1893) (figure 16D)

Diagnosis: Rectanguloid elliptical-shaped shell, with a short anterior region that is slightly lower than the posterior one. Dorsal edge regularly arched and ventral edge straight or slightly arched. Short, lamellar pseudocardinal teeth, the posterior one slightly reinforced. Periostracum dark brown.

Diplodon sp (figure 16E)

Diagnosis: Rectanguloid, equivalent shell, with the anterior region rounded, and with the umbo positioned in the first third. Straight ventral edge, with a beak in the posterior region and a thickening in the anterior region. The left valve has two long lateral teeth and a pseudocardinal tooth, serrated obliquely, which ends abruptly at the scar of the anterior adductor muscle. The right valve has one lateral tooth and two pseudocardinal teeth, the dorsal one small and the ventral one well developed. Periostracum with a golden brown color and a pearly inner surface, with iridescent yellow spots.

Castalia undosa undosa (Martens, 1827) (figure 16F)

Diagnosis: Subtriangular, equidistant and inequilateral shell, with a prominent carina that is truncated in the posterior region. Lúnula easily visible. Well-developed scutellum, with two cardinal teeth on each valve. The anterior cardinal tooth is larger than the posterior one, with several well-developed cusps. The inner and outer surfaces of the cardinal tooth on the left valve are crenulated, while on the right valve the crenulations are only present on the inner surface. In the right valve there is a lateral tooth with both surfaces showing crenulations, and in the left valve there is a side of equal size whose inner surface is crenulated.

Anodontites trapezeus (Spix, 1827) (figure 16G)

Diagnosis: Subcircular, equilateral shell with almost straight dorsal and rear edges. Two non-prominent ribs run from the umbo towards the posterior region, the dorsalmost of which ends in the region of the diaphragm, between the two openings. The valves are inflated, touching along the entire outer edge, and the junction has no teeth. The umbo is distended from above the junction line.

Fossula fossiculifera (Orbigny, 1835) (figure 16H)

Diagnosis: Shell with a subcircular, equivalent and inequilateral contour, with a winged posterior edge starting above the umbo. Two slightly marked ribs originate at the umbo and extend to the ventral region, with the more dorsally positioned one ending in the region of the diaphragm. Well-developed tail, with one pseudocardinal tooth on the left valve and two pseudocardinal teeth on the right valve. Shallow umbonal cavity.

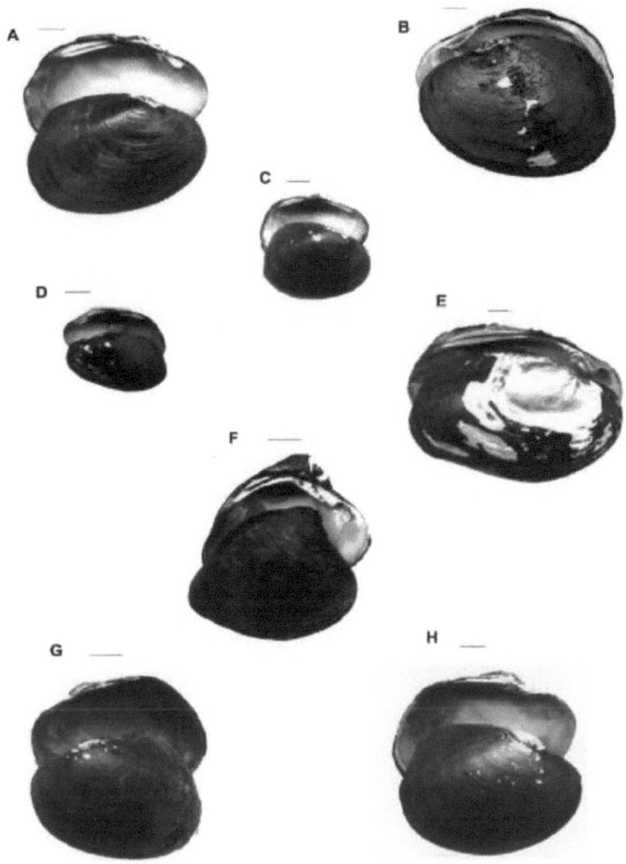

Figure 16. Species collected in the Mogi Guaçu River, Porto Ferreira, SP (21°50'36.1"S and 47°29'44.5" W). A, Diplodon expansus; B, Diplodon rotundus gratus; C, Diplodon fontainianus; D, Diplodon martensi; E, Diplodon sp; F, Castalia undosa undosa; G, Anodontites trapezeus; ***H****, Fossula fossiculifera. Scale = 1 cm.*

Table 4: Native limnic Bivalves collected from the Mogi Guaçu River, municipality of Porto Ferreira, Sao Paulo, Brazil, between March/2008 and September/2009.

Order	Family	Species	Quantity
		Diplodon expansus	129
		Diplodon martensi	5
		Diplodon fontainianus	27

	Hyriidae	*Diplodon rotundus gratus*	13
Unionoida		*Diplodon* sp	2
		Castalia undosa undosa	286
		Trapezoid anodontitis	27
	Mycetopodidae	*Fossula fossiculifera*	21

It should be noted that in some regions of the study area *Corbicula fluminea,* an exotic species, represented almost 90% of the bivalves collected. Despite this, the structure of the native community can be described as 90.6% belonging to the Hyriidae family, mainly *C. undosa undosa* (61.9%) and *D. expansus* (27.9%). Among the Mycetopodidae collected, a numerical balance was found between the species.

In our collections, we observed that the representatives of the *D. expansus* species showed great shell plasticity, making it possible to differentiate between two groups, with their intermediate shapes (figure 17).

Figure 17. Photo of the variation in the shells of D. expansus, collected in the Mogi Guaçu river, municipality of Porto Ferreira, Sao Paulo.

For identification we use both conquiliological and anatomical features.

The conquiliological characters that defined a group were:

The shell is equivalent, inequilateral, with a straight dorsal margin, oval in the anterior region and sub-square in the posterior region, with a strong amphidelic ligament. Flattened (platycotic) valves. Umbus Located in the first third of the anterior region, it is not very prominent and the periostracum is worn away. With a narrow lunula, and in larger specimens the lunula is wider. The periostracum is brownish-brown in color, with shades varying from brown to black as the shells get larger. The inner surface is smooth and whitish. The pallial line is well marked, starting in the posteroventral region of the posterior adductor muscle and ending in the ventral region of the anterior adductor muscle.

From the umbo in a posterior direction to the margin of the shell there is a shallow groove which widens to the region of the anal opening. There is a rib from the posterior side of the umbo to the edge of the shell in the region of the diaphragm. And a second, more subtle ornamentation, above the first, up to the median region of the shell. Growth lines concentric to the umbo, ending in a straight line in the anterior region.

Short, delicate scabbard, posterior to the umbo, of the schizodont type. The length of the ligament runs from the ligamental isthmus anteriorly to the end of the pseudocardinal tooth.

The left valve has a lamellar pseudocardinal tooth, striated obliquely in an anterior direction, both externally and internally. The second pseudocardinal tooth is underdeveloped, with several denticles on the edge. But never more than half of the first, more dorsal pseudocardinal tooth. And two lateral lamellar teeth. The inner surface of the ventral lateral tooth is smooth and the surface facing the interdentum region is grooved. Dorsal lateral tooth with smooth inner and outer surfaces, smaller than the ventral lateral tooth.

The right valve has two pseudocardinal teeth. A dorsal one, with a lamellar structure, striated obliquely in an anterior direction, with a smooth outer surface and small crenulations on the edge. The ventral tooth is also obliquely striated in an anterior direction, with a smooth outer surface and a crenulated edge. This tooth is thicker than the dorsal pseudocardinal tooth. The lateral tooth starts at the umbo posteriorly and runs to the dorsal region of the posterior adductor muscle. It is slightly striated in the upper terminal portion, with a smooth outer surface and a rough inner surface in the terminal portion (edge of the tooth).

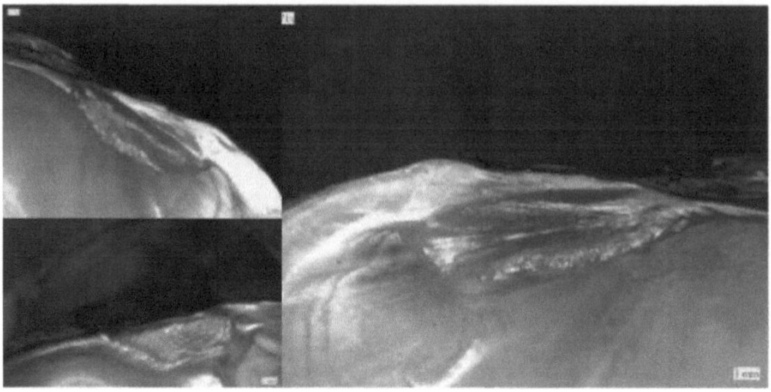

Figure 18. Details of the pseudocardinal teeth of the valves of group 1 of Diplodon expansus. On the left the pseudocardinal tooth of the left valve, and on the right the pseudocardinal tooth of the right valve.

The second group was defined by specimens with:

The shell is equivalent, inequilateral, with a rounded anterior region and a sub-square posterior region, presenting a strong amphidellic ligament. Dorsal region practically straight and with the posterior region wider than the anterior. Valves a little inflated, with the umbonal region not very high, usually eroded. Lúnula, not always observed. Periostracum dark brown and smooth, whitish nacreous layer with bluish rays. Thick shell. On the outer surface of the valves there is a groove starting at the umbo towards the posteroventral region, ending near the inhalant and exhalant openings, the growth lines are well marked concentric.

Left valve with a pseudocardinal tooth with striations, coarser on the inner side, and a second pseudocardinal tooth almost vestigial.

The pseudocardinal tooth begins just after the umbo and ends above the anterior adductor muscle. It has two lamellar lateral teeth. The inner lateral tooth is externally smooth and internally striated only in the terminal region. The outer lateral tooth is smooth both externally and internally. The inner lateral tooth is slightly larger than the outer one, leaving only the terminal portion showing its striations, it is longer but the same height as the outer tooth.

Right valve with two pseudocardinal teeth and a lateral tooth. The dorsal pseudocardinal tooth is short and robust, with grooves on both sides. The ventral one is lamellar, striated obliquely in an anterior direction on the side facing the interdentum, the other side is smooth.

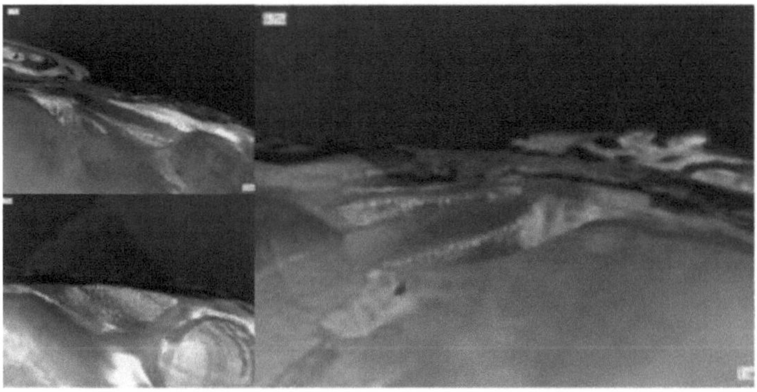

Figure 19. Details of the pseudocardinal teeth of the valves of group 2 of Diplodon expansus. On the left the pseudocardinal tooth of the left valve, and on the right the pseudocardinal tooth of the right valve.

When the anatomical characters were used and compared with the conquiliological ones, it was possible to identify a third group.

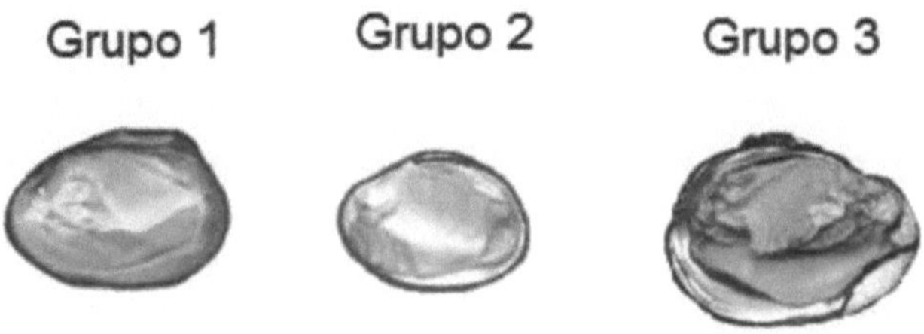

Figure 20. Anatomical features of the three groups of Diplodon expansus.

Anatomy group 1:

Slit-shaped exhalation opening, with its upper portion closed. Exhalant opening as long as the inhalant, separated from it by a solid mantle connection. Inhalant opening with distinct papillae. Palpus moderate, subtriangular, posterior margin connected for about a third of its length.

Subtrapezoidal inner demibranch, taller and wider than the outer one, with its anterior portion connected to the palpus and its ventral margin rounded. Subtriangular external demibranch, occupying 4/6 of the internal demibranch. Posterior portion more elongated than the inner demibranch and anterior region well tapered, leaving the inner demibranch fully exposed. Internal lamella of the internal gill fully connected to the abdominal sac.

Anatomy group 2:

Thin mantle, well pigmented at the edge, salmon orange, pigmentation more evident in the region of the openings. The edge of the aperture region is dark brown tending to black, and the edge of the mantle is thicker than the middle. Inhalant aperture entirely open, with small papillae, exhalant aperture separated from inhalant by a strong, black diaphragm.

The palpus is rounded in the anterior region and slightly tapered in the posterior region, with very strong orange pigmentation. The outer part of the palpus is pleated along the edge from the anterior to the posterior region, the dorsal part of the palpus is not pleated, only the ventral part is pleated. The entire inner part of the palpus is pleated, not just the edge. The folds on the inside of the palpus run from the mouth opening to the posterior edge of the palpus. The pigmentation is on the outside of the palpus, the dorsal edge of the palpus is slightly thicker than the ventral edge of the palpus.

Internal and external demibranchs with a subtrapezoidal shape, both with a straight ventral margin. External demibranch slightly smaller in the anterior region, but with the same height in the posterior region, completely covering the internal one.

Anatomy group 3:

Thin mantle, not very pigmented, with a thickening at the edge. Inhalant and exhalant apertures of similar size, but the inhalant has a series of papillae in the region near the diaphragm. Dark-colored foot.

Subtriangular palpus with folds on the edge, longer than wide. Internal and external demibranchs subtriangular, with the same height. External demibranch slightly larger in the posterior region, with a sort of fringe on its posteroventral portion.

The specimens were then named *D. expansus* var. 1, *D. expansus* var. 2 and *D. expansus* var. 3, for identification purposes for the propagation experiments.

3. Bivalve propagation in the laboratory

3.1. Isolation, cultivation and maintenance of microalgae

Among the various species of algae collected, the most representative species and the one that developed most strongly in the laboratory was *Chlamydomonas* SP (figure 21). It was well accepted by the bivalves and was more resistant to bacterial and/or protozoan contamination in the cultures, which is why it was selected for mass production.

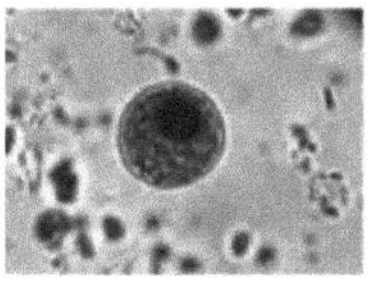

Figure 21. Species of microalgae isolated from the Mogi Guaçu river and cultivated on a large scale at the

The culture medium with inorganic fertilizer at a concentration of 20/200 reached its highest cell density on the fourth day, while all the other treatments showed a population peak between the seventh and eighth day ($p < 0.05$) (figure 22).

The highest cell densities were observed in cultures with inorganic fertilizer (in the proportions 1/200 and 5/200) and cow manure, with values of 1230, 1170 and 1380 x 10^4 cells/ml, respectively, showing no significant difference between them ($p < 0.05$).

The other concentrations, 0.5/200, 10/200 and 20/200, allowed the population density of the cultures to reach lower values of 710, 420 and 239 x 10^4 cells/ml, respectively.

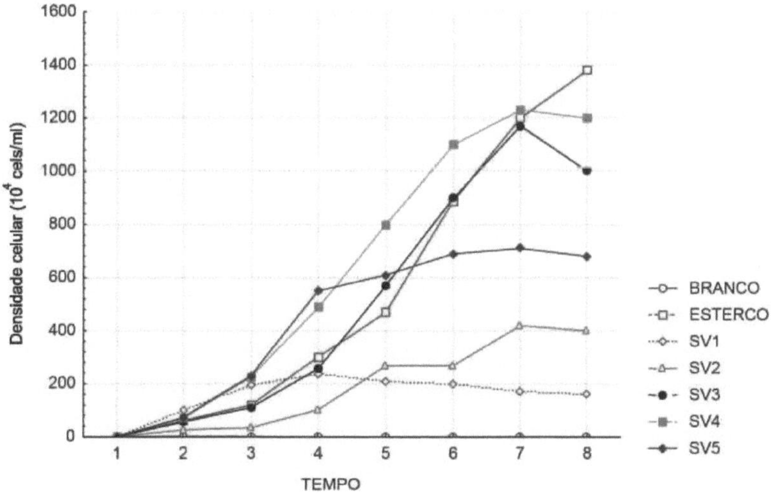

Figure 22. Growth in population density of Chlamydomonas sp cultivated in different culture media.

Both the growing medium and the procedures used in the production of microalgae led to the production of 500 liters of food per month.

3.2. Keeping adult bivalves in the laboratory

3.2.1. Preliminary experiments

The group of animals kept suspended showed greater survival than those buried ($p < 0.05$), with only two animals (4%) dying over the 120-day period. The buried bivalves ended the experimental period with 22% mortality (11 animals).

Despite the random allocation, the buried animals were, on average, heavier than the suspended ones from the beginning of the observation period (average difference of 26.8 g, I.C. (95%) = [11.5; 42.1], p = 0.0005).

In the ANOVA model, the interaction effect was significant ($p < 0.001$), indicating that the growth of the animals was different in the two groups. In the group of suspended animals, there was a weight loss in the 1st month ($p < 0.001$) and from then on the weight gain was gradual with all differences statistically significant ($p < 0.001$). In the buried group, there was a significant weight loss in the 1st and 2nd month ($p < 0.001$), but there was no difference between the 2nd and 3rd month ($p = 0.1226$), nor between the 3rd and 4th month ($p = 0.6963$). At all assessment times, the average weight in the buried group was higher than in the suspended group ($p < 0.01$) (table 5).

Table 5. Weight, in grams (means and standard deviations), of specimens of Anodontites trapesialis kept buried and suspended.

Evaluation moment	Initial	1st month	2nd month	3rd month	4th month
Buried	140,1	138,5	133,4	136,2	134,2
	(44,3)	(44,4)	(41,0)	(41,2)	(40,5)
Suspended	113,3	111,6	113,5	114,4	115,7
	(31,8)	(31,9)	(32,2)	(32,1)	(31,8)

In summary, both groups lose weight in the first month. The suspended animals manage to gradually regain weight, while the buried animals not only lose more, but their weight remains low (lower than their initial weight).

In percentage terms, comparing the weights measured at the beginning and end of the observation period, there was an average gain of 2.1% in the suspended group and an average loss of 1.4% in the buried group.

3.2.2. Maintenance of collected bivalves

Both the methodology used and the diet provided by the microalgae species cultivated proved to be efficient for maintaining the bivalve specimens kept in the laboratory. The weight of the animals remained constant during the period in which they were fed the diet offered (*Chlamydomonas* sp and *Ankistrodesmus* sp). It was possible to keep *C. undosa undosa*, *D. rotundus gratus*, *D. expansus*,

D. fontainianus, D. martensi, F. fossiculifera and *A. trapezeus*, with a mortality rate of less than 5%.

It was observed that the species *C. undosa undosa* showed greater survival with a more constant flow of water, while the species belonging to the Mycetopodidae family, *F. fossiculifera* and *A. trapezeus*, did not show a difference in survival in still water conditions.

3.3. Obtaining gloquidium

3.3.1. Glochidia release

Of the individuals collected from *D. fontainianus*, only 12 animals (44.44%) released gloquidia (figure 23A) in the laboratory, and the highest number of releases was recorded in August (table 6).

Two specimens of *D. martensi* (40%) naturally released their gloquidia (figure 23B) in September (table 6).

46.15% of the collected specimens of *D. rotundus gratus* (figure 23C) released their gloquidia in June and July (table 6).

As it was only possible to identify the varieties of *D. expansus* by observing the internal anatomy, it is difficult to pinpoint the percentage of individuals that released their glochidia in the laboratory. As it was only possible to present the absolute number of animals that released gloquidia, it was observed that 7, 14 and 8 specimens of *D. expansus* var. 1, 2 and 3 released their gloquidia (figure 23D) between July and September (table 6).

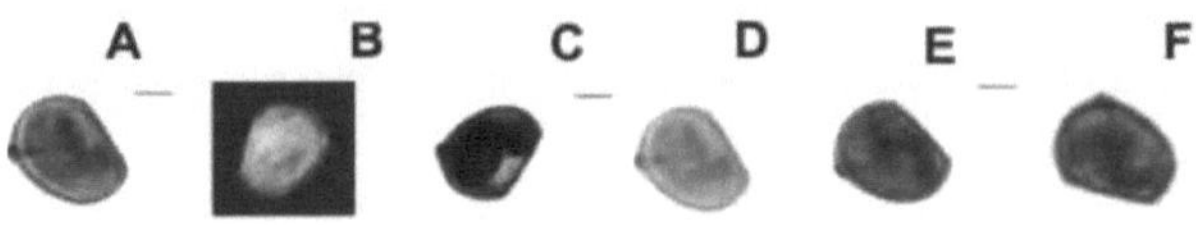

Figure 23. Glochidia of the species of the genus Diplodon that were liberated in the laboratory. Legend: A, Diplodon fontainianus; B, Diplodon martensi; C, Diplodon rotundus gratus; D, Diplodon expansus var. 1; E, Diplodon expansus var. 2; F, Diplodon expansus var. 3. Scale= 100 µ.

26.57% of the *C. undosa undosa* individuals collected were released between October and November (Table 7).

The species of *Diplodon* whose gloquidia were ready to be released or were in the process of being released from the marsupials began to be eliminated between 12 and 48 hours after being stored in the laboratory. The way in which the species release the gloquids showed different behavior; *D. rotundus gratus* release is continuous and occurs through the exhalant opening in the form of a mucous ribbon lasting around 10 to 12 hours. In *D. expansus*, release occurs, also via the exhalant opening, in regular pulses lasting more than 24 hours.

Table 6. Species collected in the Mogi Guaçu River, in the municipality of Porto Ferreira (21°50'36.10" S and 47°29'44.5" W) and number of individuals that spawned in the laboratory.

Species	Collected	Spawned
D. expansus var. 1		7
D. expansus var. 2	129	14
D. expansus var. 3		8
D. fontainianus	27	12
D. martensi	5	2
D. rotundus gratus	13	6
C. undosa undosa	286	76
A. trapezeus	27	3
F. fossiculifera	21	10

3.3.2. Viability of gloquidia

The viability of the gloquidia *of D. expansus* var. 2 and *D. rotundus gratus* over time was similar and consistent. For the comparison of treatments and species, the value of 75% viability was selected as the lower threshold. Although arbitrary, lower values, according to Zimmerman and Neves (2002), would negate any benefit when compared to the *in situ* viability of gloquids. In all cases, the treatments subjected to lower temperatures resulted in greater survival than those subjected to higher temperatures (tables 7 and 8).

The treatments with distilled and deionized water showed no significant differences ($p < 0.05$) between the temperatures tested with *D. expansus* (table 7). In the experiments with *D. rotundus gratus*, only at 18 °C did the use of deionized water lead to greater viability of the larvae ($p < 0.05$) (table 9).

The use of an antibiotic/fungicide solution after the washing process meant that the viability of the gloquids lasted for a longer period of time. And that washing the gloquids beforehand prevented fungus from occurring during the incubation period, thus increasing the survival of the larvae.

Those gloquids that did not undergo any kind of process (with mucus) showed low viability due to strong bacterial contamination in the first 24 hours. Another treatment that proved ineffective was the use of a 50 µm screen, which caused total mortality of the gloquids, probably due to the mechanical shock the larvae suffered during the process.

Table 7. Number of hours that the viability of D. expansus var. 2 gloquidia remained above 75%. Periods within a row followed by different letters were significantly different (p < 0.05).

Temperatures (°C)			
Treatment of gloquidia	18	22	24
With mucus	32^{a}	28^{a}	26^{a}
Autoclaved mine water	148^{a}	96^{b}	72^{b}
Distilled water	175^{a}	132^{b}	108^{c}
Deionized water	178^{a}	140^{b}	108^{c}
50 µm screen	48^{a}	44^{a}	44^{a}
Antibiotic/antimycotic solution	196^{a}	175^{b}	157^{c}

Table 8. Number of hours that the viability of D. rotundus gratus gloquidia remained above 75%. Periods within a row followed by different letters were significantly different (p < 0.05).

Temperatures (°C)			
Treatment of gloquidia	18	22	24
With mucus	32^{a}	28^{a}	26^{a}
Autoclaved mine water	122^{a}	96^{a}	64^{c}
Distilled water	158^{a}	128^{b}	100^{c}
Deionized water	164^{a}	130^{b}	104^{c}
50 µm screen	52^{a}	48^{a}	44^{a}
Antibiotic/antimycotic solution	187^{a}	146^{b}	122^{b}

3.3.3. Gloquid conchylometry

The glochidium of *D. rotundus gratus* showed the highest mean values of the measurements taken, with the exception of the angle of obliquity, with values of length of 294.2 µm, height of 240.35 µm, length of the dorsal line of 215.35 µm, displacement of the ventral tip of 38.47 µm and an angle of obliquity of 16°. The glochidia *of D. expansus* var.1 and var.2 showed average values for length of 279.72 µm and 276.46 µm, height of 224.76 µm and 226.6 µm, dorsal line length of 192.55 µm and 193.3 µm, ventral tip displacement of 26.55 µm and 18.3 µm, and angle of obliquity of 18.41° and

18.5°, respectively (table 9).

Table 9. Morphometric measurements of the gloquid shells of D. expansus var. 1, D. expansus var. 2 and D. rotundus gratus.

Diplodon expansus var. 1

n = 40	**variation**	**fashion**	**average**	**detour standard**	**hp %**
Length	263,3 - 300	273,3	279,72	1,65	0,5898
Height	206,6-253,3	226,6	224,76	3,3	1,4682
CDC	140 - 220	193,3	192,55	13,35	6,9333
DPV	6,6 - 66,6	13,3	26,55	13,35	50,2872
A°	12 - 23,5	17	18,41	2,5	13,5777

Diplodon expansus var. 2

n = 40	**variation**	**fashion**	**average**	**detour standard**	**hp %**
Length	260 - 293,3	280	276,46	7,56	2,7359
Height	210 - 246,6	226,6	226,6	0	0
CDC	180 - 213,3	200	193,3	6,7	3,4661
DPV	6,6- 60	26,6	18,3	1,7	9,2896
A°	10 - 23	18	18,5	2,96	16,0161

Diplodon rotundus gratus

n = 40	**variation**	**fashion**	**average**	**detour standard**	**hp %**
Length	276,9-311,5	290,4	294,83	16,91	5,7367
Height	226,9-253,8	238,5	243,07	3,07	1,2650
CDC	176,9-253,8	207,7	219,24	3,86	1,7606
DPV	15,4 - 61,54	30,8	42,31	11,54	27,2749
A°	12 - 20	15	16,25	2,85	12,9161

Legend: **CDC**, dorsal shell length; **DPV**, ventral point displacement; **A°**, angle of obliquity; **cv**, coefficient of variation.

3.4. In vivo cultivation of gloquidia

Among the fish species collected in the Mogi Guaçu river, *Geophagus brasiliensis, Hyphessobrycon eques, Cheirodon* sp and *Papiliochrommis ramirezi* did not prove to be hosts for the naiads of the species *D. rotundus gratus, D. expansus* var. 1, *D. expansus* var. 2 and *D. martensi*, since, regardless of the temperature used, they did not have any gloquidia attached to their bodies. The same occurred with the exotic species *Oreochromis niloticus*.

The species *Gymnotos* sp and *Randia queli*, when exposed to the gloquidia *of D. expansus* var. 2 at 24 °C, showed cysts 24 hours after artificial infestation. The gloquidia adhered to the caudal fin and barbels of *Gymnotos* sp and *Randia queli*, respectively. The first species mentioned, after five days, tore its own fin with the cysts, eliminating the gloquidia. The second species, on the other hand, showed some kind of immune response and the gloquidia detached themselves from the barbels after four days.

The species that proved to be potential hosts for the species of the genus Diplodon collected and tested were *Astianax altparanae* and *Poecilia* sp. The latter, because of its very small size (1-2.5 cm), was unable to support the development of the larvae until metamorphosis, perishing after 7-10 days.

The percentage of metamorphosed gloquidia of the different species tested and their incubation time, when brought into contact with *A. altparanae*, can be seen in Table 10.

Table 10. Duration of the infestation period and percentage of metamorphosis, in Astyanax altparanae, according to the species of bivalve and the temperature of the experiment.

Bivalve species	Temperature (°C)	Infestation period (days)	Metamorphosis (%)
	20	15	5,3
D. rotundus gratus	22	13	3,6
	22	12	1,2
D. expansus var. 1	24	9	0,8
	20	15	1,8

D. expansus var. 2	24	10	1,0
D. martensi	20	18	2,4

3.5. *In vitro cultivation of gloquidia*

3.5.1. Culture medium (M 199 + antibiotic/fungicide)

The gloquids that were cultivated with the culture medium containing the HEPES buffer solution died suddenly during their development (results in the tables on gloquid cultivation, sub-item 3.5.3.), regardless of the species, temperature or protein source used. With the removal of the above-mentioned solution, the *in vitro* cultivation of limnetic bivalves of the genus Diplodon was successful.

The use of the antibiotic/fungicide combination did not prevent the formation of fungal colonies and the appearance of protozoa in the cultures, even when only the medium was used, without the inclusion of fish plasma (figure 24).

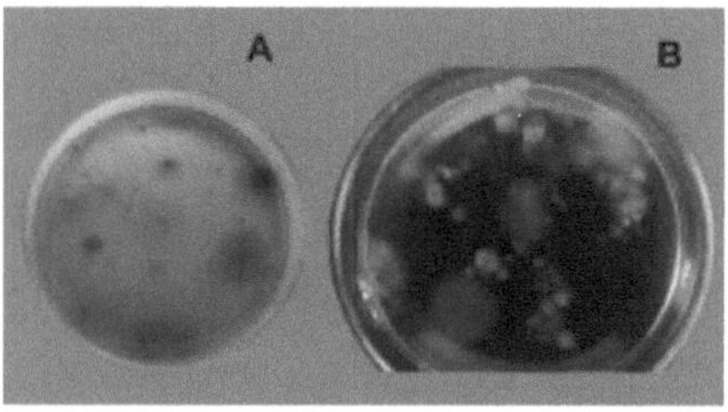

Figure 24. Photos of the contaminated gloquid cultures. A, incubation with culture medium only; B, incubation with culture medium and fish plasma.

3.5.2. Protein source

3.5.2.1. Fish plasma

The extraction of blood from the tail vein was satisfactory, since it was possible to obtain blood from large and small species (e.g. *Colossoma* sp and *A. altparanae*, respectively) and at the same time keep the fish alive to use them again. An average of 7 ml of blood/kg of fish was obtained. The percentage of plasma after centrifuging the blood varied between 65 and 30%, depending on the species used.

3.5.2.2. Freeze-dried fish extract

The preparation of freeze-dried fish extract is extremely simple, and the maintenance/preservation of this extract is much less laborious when compared to the preparation and preservation of fish plasma.

3.5.2.3. Bovine plasma

The bovine blood was obtained by the employees of the slaughterhouse and the aseptic conditions were not controlled. After obtaining the plasma, it proved to be viable (in terms of contamination or coagulation) for a very short period of 15 days, even though it was frozen.

3.5.3. Growing gloquidia

The gloquidia incubated at both 24 °C and 20 °C showed strong bacterial contamination, with total mortality on the fourth day of cultivation. When the gloquidia were incubated at 18 °C, they could be cultivated until metamorphosis.

The use of plasma from *Colossoma* sp, *Oreochromis* sp and *Cyprinus* sp as a protein source for the culture medium allowed the gloquidia *of D. fontainianus* and *D. expansus* var. 2 to develop for a maximum of six days, with total mortality occurring after this period (Table 11). When *A. altparanae* plasma, freeze-dried *A. altparanae* extract and bovine plasma were used as protein sources, the gloquidia of *Diplodon martensi*, *D. rotundus gratus* and *D. expansus* var. 2 developed until metamorphosis (table 12). When using the freeze-dried extract of *A. altparanae*, the survival and metamorphosis rates for *D. rotundus gratus* were 40% and 20% respectively. For *D. expansus* var. 2, the survival rate was 50% with 15% of individuals metamorphosing.

In the glochidia *of D. rotundus gratus* and *D. expansus* var. 2, cell extrusion was noticed from day 12, which remained until day 20, and was later replaced by the appearance of the mantle. In the glochids adhered to the host fish, cell extrusion was also observed, but only 1-2 days after metamorphosis. The appearance of gill arches was noted *in the* gloquids cultured *in vitro* on days 19-21. And the gloquidia showed a ciliated foot on day 22 of cultivation, indicating metamorphosis to the juvenile stage.

Table 11. Survival time of Diplodon fontainianus and D. expansus var. 2 gloquidia incubated at 18 °C with plasma from different fish species as protein sources in the culture medium.

Plasma of the species	***D. fontainianus***	***D. expansus* var. 2**
	Survival time (Hours)	**Survival time (Hours)**
Colossoma sp	*36*	*36*
Oreochromis sp	144	120
Cyprinus sp	144	120

A. altparanae	nt	528

Caption: **nt**, not tested.

Table 12. Survival percentage (and metamorphosis percentage) of Diplodon rotundus gratus, D. expansus var. 2 and D. martensi gloquids incubated at 18 °C with different protein sources in the culture medium.

Protein source	***D. rotundus* *gratus***	***D. expansus* var. 2**	***D. martensi***
Bovine plasma	*1 (0,1)*	*1 (0,08)*	nt
Lyophilized extract	40 (20)	50 (15)	40 (35)
Plasma A. altparanae	nt	50 (5)	nt

Caption: **nt**, not tested.

3.6. Growing juveniles

The cultivated species, *D. rotundus gratus*, *D. expansus* var. 1 and var. 2, and *D. martensi*, showed the same behavior during their growth, thus allowing the creation of a protocol for the cultivation of juveniles of the genus Diplodon. The need to use different rearing systems for each stage of development was verified, and this was divided into rearing post-metamorphic juveniles, five-day-old juveniles, 15-day-old juveniles and juveniles older than 30 days. Tables with the survival results of the juveniles subjected to the various treatments were drawn up using the survival averages of the different species tested.

3.6.1. Post-metamorphic juveniles (figure 25)

The post-metamorphic juveniles that were kept in Petri dishes showed better survival when compared to those kept in Beckers (Table 13). When comparing the types of treatments to which the gloquids were subjected, the best results were obtained in the treatments without adding food, regardless of the container used. However, those gloquids that were placed in contact with the sediment showed significantly higher mortality.

Table 13. Survival of post-metamorphic juveniles of the genus Diplodon. The survival values represent an average between the species tested.

	Petri dish				**Becker**			
DAY	**SALT**	**MIC**	**SED**	**MIC+SED**	**SALT**	**MIC**	**SED**	**MIC+SED**
0	20	20	15	15	15	15	15	15

2	19	10	10	5	13	7	10	7
4	19	10	7	0	12	5	4	0
6	15	0	3	0	10	0	2	0

Key: **SAL**, without food; **MIC**, addition of microalgae; **SED**, presence of sediment; **MIC+SED**, with food and sediment.

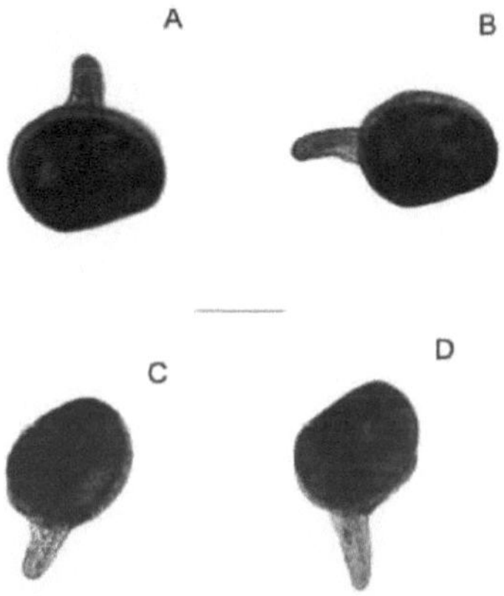

Figure 25. Photographs of post-metamorphic juveniles. A, Diplodon expansus var. 1; B, Diplodon expansus var. 2; C, Diplodon martensi; D, Diplodon rotundus gratus. Scale = 200 μ.

3.6.2. Juveniles aged 5 days (figure 26)

The juveniles that were kept in Petri dishes with the addition of 30000 cells/ml of the microalga *Chlamydomonas* sp, showed a higher survival rate, but without showing a significant difference with those that were kept in Petri dishes with the addition of sediment, maintaining a survival rate of 80%, 72% and 80% in the treatments with the addition of food, only with sediment and with the addition of food and sediment, respectively (Table 14).

Individuals older than five days that were kept in Becker with aeration showed a high mortality rate. The sediment treatment proved to be significantly more viable, with a survival rate of 66.66 % after twelve days of experimentation.

Table 14. Survival of post-metamorphic juveniles of the genus Diplodon. The survival values represent an average between the species tested.

Petri dish						**Becker**		
DAY	**SALT**	**MIC**	**SED**	**MIC+SED**	**SALT**	**MIC**	**SED**	**MIC+SED**

0	10	25	25	25	10	15	15	15
2	7	24	22	22	5	11	14	14
4	5	24	20	22	5	11	13	13
6	2	23	20	22	0	7	10	11
8	2	20	19	20	0	7	10	11
10	0	20	19	20	0	4	10	11
12	0	20	18	20	0	4	10	10

Key: **SAL**, without food; **MIC**, addition of microalgae; **SED**, presence of sediment; **MIC+SED**, with food and sediment.

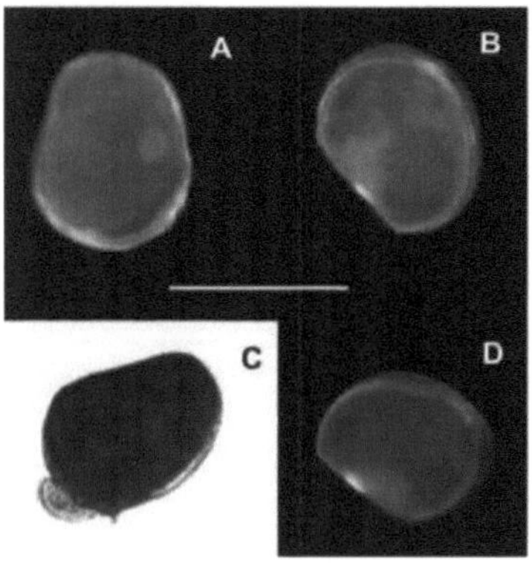

Figure 26. Photographs of juveniles. A, Diplodon expansus var. 1 at 5 days old; B, Diplodon expansus var. 2 at 5 days old; C, Diplodon martensi at 5 days old; D, Diplodon rotundus gratus at 5 days old. Photographs A, B and D were taken with epifluorescence microscopy. Scale = 500 µ.

3.6.3. Juveniles aged 15 days (figure 27)

After 15 days of rearing, the juveniles behaved differently, with greater survival among those kept in Beckers with aeration (Table 15). Among the treatments tested, the one with the best results was the one with the addition of microalgae and sediment, with a survival rate of 85 %, significantly higher than the others, where survival did not exceed 40 %.

Table 15. Survival of juveniles at 15 days after metamorphosis of the Diplodon genus. The survival values represent an average between the species tested.

Petri dish	**Becker**

DAY	SALT	MIC	SED	MIC+SED	SALT	MIC	SED	MIC+SED
0	10	15	20	20	10	15	20	20
2	10	15	20	20	9	15	18	20
4	8	15	18	17	5	11	18	19
6	3	14	14	17	2	11	17	19
8	0	11	14	17	1	11	17	17
10	0	4	8	14	1	8	11	17
12	0	4	8	14	0	8	11	17
14	0	1	8	12	0	8	8	17
16	0	1	7	5	0	8	8	17

Key: **SAL**, without food; **MIC**, addition of microalgae; **SED**, presence of sediment; **MIC+SED**, with food and sediment.

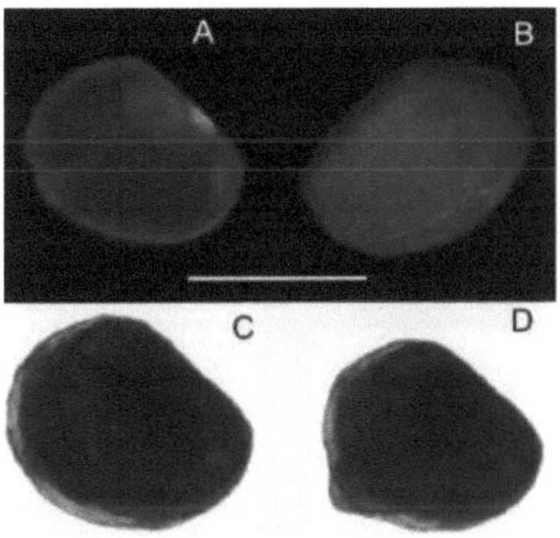

Figure 27. Photographs of juveniles. A, Diplodon expansus var. 1 at 15 days old; B, Diplodon expansus var. 2 at 10 days old; C, Diplodon martensi at 15 days old; D, Diplodon rotundus gratus at 15 days old. Photograph A was taken with epifluorescence microscopy. La = 500 µ.

3.6.4. Juveniles aged 30 days or more (figure 28)

Following the working protocol, the newly metamorphosed gloquids were kept in Petri dishes (60x15 mm) with autoclaved mine water, previously shaken with air diffusers, at a maximum density of 20 juveniles/plate. *Chlamydomonas* sp. was added every other day from the fourth day onwards. Before feeding the juveniles, 2/3 of the water in each Petri dish was changed.

On the 12th day, the juveniles were transferred to Petri dishes with a 1 mm layer of sandy sediment (particles < 125 µm) and fed every other day, changing 2/3 of the water with each addition of microalgae. On the 20th day, the juveniles were placed in a 150 ml beaker (maximum density of 50 juveniles/Becker) with 1 mm of sediment and air diffusers to keep the microalgae in the water column and oxygenate the medium. The containers were washed and the sediment changed every two weeks.

With the methodology employed, it was possible to obtain juveniles over 30 days old for the species *D. rotundus gratus*, *D. martensi* and *D. expansus* var. 2. The juveniles *of D. martensi* had a mortality rate of 65 % up to 100 days of rearing, *D. expansus* var. 2 was reared up to 65 days with a mortality rate of 40 %. The juveniles *of D. rotundus gratus* remained alive until 62 days of rearing.

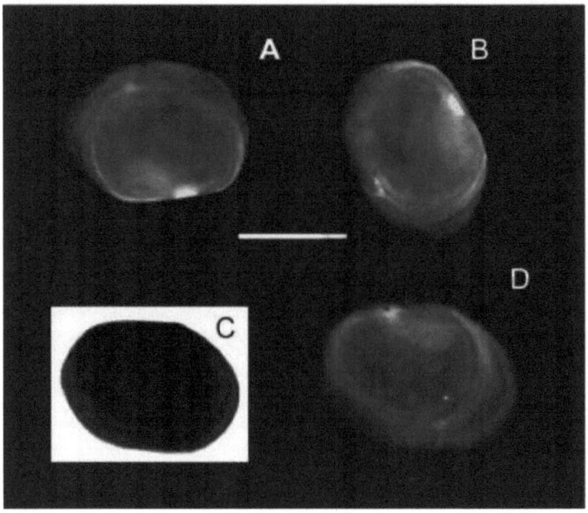

Figure 28. Photographs of juveniles. A, Diplodon rotundus gratus at 30 days; B, Diplodon martensi at 30 days; C, Diplodon martensi at 60 days; D, Diplodon martensi at 90 days. Photographs A, B and D were taken with epifluorescence microscopy. Scale = 200 µ.

Discussion

Reintroduction is a tool for recovering degraded environments (Neves and Zale, 1982; Watters, 2000). However, the ability to produce juvenile bivalves in laboratories will be of little value for biodiversity conservation if these individuals do not survive after release (Zimmerman, 2003). The key to this issue is the identification of suitable release sites. There are currently many sites where populations are extremely low or have become extinct, but which could once again support viable populations through the augmentation or reintroduction of bivalves (Henley and Neves, 1999).

Concern about depression and the loss of rare alleles has fueled opposition to the introduction of captive-bred juveniles (Hoeh *et al.*, 2009, Rogers, 1999). Furthermore, since habitat degradation and poor water quality are two of the main reasons for the initial decline, it is not certain that propagated individuals could survive and reproduce. Issues of conservation genetics and habitat quality, however, must be weighed against the urgency of recovery and the threat of extinction. Reintroductions of species to historical sites can increase the number of healthy populations and thus reduce the possibility of species extinction in the face of a stochastic event (McMahon and Bogan, 2001). Long-term restoration objectives should include protection measures to prevent population decline, as well as measures that include habitat restoration, the variety of fish species that co-habit with bivalve populations and the propagation of juveniles for reintroduction (Zimmerman, 2003).

The propagation of naiads in the United States began in the early 1900s using techniques very similar to those used today. This experience provided scientists with a basis for understanding at least the basic requirements of this complex group of invertebrates. Propagation efforts are currently limited to *in vivo* techniques, which include the host fish (Lefevre and Curtis, 1912) and another, *in vitro*, using artificial culture media (Isom and Hudson, 1982). However, the use of artificial culture media may be the only way for some bivalves to survive extinction (Ahlstedt, 1981; Neves *et al.*, 1997). Understanding the life history of this unique group of invertebrates also requires us to analyze the life history, physiology, biochemistry and immunochemistry of the host fish. This understanding of the physiology of the host fish was a prerequisite for the development of artificial culture media (*in vitro* technique) in the early 1980s, providing information on the specific nutrient requirements to support the development and growth of glochids (Farris and Van Hassel, 2006).

Techniques for determining fish hosts have been reported and used by various researchers over the decades (Howard, 1915; Coker *et al.*, 1921; Penn, 1939; Cope, 1959; Giusti *et al*, 1975; Jenkinson, 1982; Hove and Neves, 1991; Hoggarth, 1992; Watters and O'Dee, 1996; Haag and Warren, 2003), while some unconventional hosts (e.g. amphibians) have also been identified as supporting the metamorphosis of gloquids (Seshaiya, 1969; Watters and O'dee, 1998). Some species have shown a relatively good degree of success in using host fish techniques, while others have proved more difficult (Jones *et* al., 2006).

By infesting the fish species collected in the Mogi Guaçu River, *Geophagus brasiliensis*, *Hyphessobrycon eques*, *Cheirodon* sp and *Papiliochrommis ramirezi*, no juveniles *of Diplodon rotundus gratus*, *D. expansus* var. 1, *D. expansus* var. 2 and *D. martensi* were obtained, as the cysts were detached from the fish before they completed metamorphosis. This result may indicate that there is specificity between the gloquid and the fish.

Zale and Neves (1982), Kat (1984), Watters and O'Dee (1996) and O'Connel and Neves (1999) stated that the specificity of the gloquid/fish association varies with the species of bivalve. Thus, if a gloquid attaches itself to a non-host fish, an abnormal cyst forms and falls off after a few days of infestation. This process is attributed to a natural immunity of the non-host fish to the bivalve species (Arey, 1932; Meyers *et al.*, 1980; Waller and Mitchell, 1989; Kirk and Layzer; 1997). However, when Dudgeon and Morton (1984) studied the intensity of parasitism in *Anodonta woodiana* larvae, they found a higher incidence in exotic fish than in native species.

Techniques for determining the viability of host fish include the use of aerated tanks, direct placement on the gill, as well as the use of anesthetics to reduce stress on the fish (Zale and Neves, 1982; Henley and Neves, 1999; Zimmerman, 2003). Although modifications to these techniques have been reported by various researchers, the fundamental approach is the same. Tanks with aeration are often used when there are viable gloquids, and a variety of fish species and age ranges are used. However, if the gills are limited and/or the fish are smaller or have small gills, direct placement on the gills using pipettes is a viable alternative to air tank techniques for fixing the gills. Anesthetizing the fish before encystment is uncertain, as the possible effects of the anesthetic may inadvertently damage the adhesion of the gloquids and their subsequent metamorphosis.

The encystment of gloquids can vary from a few days to several months, depending on the species of bivalve, the health of the fish (whether the individual is stressed or sick due to other environmental factors), the temperature of the water, and perhaps other variables that are currently unknown (Mansur, 1999; Zimmerman and Neves, 2002; Zimmerman, 2003).

In our study, the glochidia *of D. rotundus gratus and D. martensi, D. expansus* var.1 and *D. expansus* var. 2 remained encysted for 9-18 days in *Astyanax altparanae* (table 11). *D. charruanus*, according to Bonetto (1954), remained encysted for a period of 10 to 20 days in *Hoplias malabaricus*. However,

Bonetto and Ezcurra (1963) found a longer parasitic period for *D. delodontus delodontus*, from 25 to 30 days. Mansur (1999) infested specimens of *Gymnogeophagus gymnogenys* with *D. martensi* gloquidia, which remained encysted for 30 days.

It was possible to observe that there is a relationship between temperature and incubation period, so at lower temperatures, the infestation period is relatively longer (Table 11). In the experiments by Bonetto (1954) and Bonetto and Ezcurra (1963) there was no mention of the temperatures at which the infestations were carried out. Dudgeon and Morton (1984) carried out infestations with *Anodonta woodiana* gloquidia and found infestation periods ranging from 14 to 25 days at 15 °C; from 8 to 12 days at 22 °C; and from 6 to 11 days at 27 °C. Tompa (1979) infested *Poecilia reticulata* with *Lasmigona compressa* larvae at 20 °C and obtained a parasitism period of 10 to 12 days. Zale and Neves (1982) used *Villosa nebulosa* gloquidia at 24.7 °C and found an infestation period ranging from 10 to 21 days. The longest parasitism period, 40 to 50 days, was obtained by Telda and Fernando (1969) who used *Lampsilis radiata siliquoidea* larvae at 15°C.

Alternatively, fish survival can be compromised by excessive gloquid infestation resulting in the limitation of gas exchange through the gill lamellae. While a ratio of 50-100 gloquids per gill for fish 15-25 cm long has been reported as adequate (Hove *et al.*, 2000), in our study we used a ratio of 200 gloquids per fish (with an average size of 5 cm). However, other authors have infested the host fish with thousands of gloquids and succeeded in metamorphosis and maintaining the viability of the fish once removed from the ponds (Milam *et al.*, 2000; Winterringer, 2004).

The number of metamorphosed individuals using *in vivo* rearing techniques is usually very low. In the present study, through infestations of various species of the genus Diplodon on *A. altparanae*, a metamorphosis percentage of only 5.3 to 0.8 % was obtained (table 11). Bigham (2002), studying the species *Venustaconcha ellipsiformis*, *Etheostoma spectabile* and *E. caeruleum*, reported an average metamorphosis rate of less than 5%. In another study carried out by Gray and colleagues (2002) with *Strophitus undulatus*, using various host fish species, the percentage of metamorphosis obtained after a 10-day incubation period ranged from 2 to 51%.

In a way, the *in vivo* cultivation of bivalves tends to imitate nature. However, this involves a great deal of labor and cost in order to carry out viability tests on potential host fish (Hove *et al.*, 2000), as well as knowledge of them (Henley and Neves, 1999). However, both in the wild and in the laboratory, there is a high mortality rate caused by bacteria, fungi and protozoa (Young and Williams, 1984a, b), which has led researchers to look for another alternative, artificial or *in vitro* culture.

Culturing gloquids *in vitro* has proven to be very efficient when compared to *in vivo* cultures. In this study, *D. rotundus gratus* produced four times more metamorphosed juveniles and *D. expansus* var. 2 7.5 times more when cultured with fish lyophilizate compared to *in vivo* culture. Similar studies have produced around ten to twenty times more juveniles from various artificial culture media (Hudson and Isom, 1984; Hudson and Shelbourne, 1990).

The interest in increasing the production of bivalves through the use of artificial culture was seen at the beginning of the 20th century, when Ellis and Ellis (1926) reported the success of the first culture for the transformation of gloquids after excising them from the gill tissue of their host fish.

Unfortunately, the details of their solutions have never been published (Farris and Van Hassel, 2006). Much later, in the early 1980s, interest in the artificial culture of gloquids was reviewed by Isom and Hudson (1982), who reported success in transforming various species without the use of a host. This technique began as a modification of modern cell culture techniques and made use of a mixture of amino acids, vitamins and glucose in Unionideo Ringer's solution (Ellis *et al.*, 1930), along with the addition of fish plasma as a source of protein, as well as growth stimulants, hormones, etc. Although this work started by mixing these components from scratch (using the concentrations found in fish plasma as a guide), Hudson and Isom (1984) also reported success using premixes, commercially available as cell culture media (Eagles' solution with essential and non-essential amino acids and M 199), which contain almost all of these amino acids in concentrations as high as or higher than those found in fish plasma.

This mixture has been used to produce thousands of juvenile bivalves (Johnson *et al.*, 1993; Hudson *et al.*, 1994; Hudson *et al.*, 1996; Barfield *et al.*, 1997; Clem, 1998), but it is not very convenient for use in some laboratories due to the requirement for a ready supply of fish so that the blood can be extracted and separated into plasma and non-plasma components. Because of this inconvenience and because the use of fish plasma shows great variation in results due to the conditions of the fish, Hudson and Shelbourne (1990) tried to develop an alternative medium that was serum/plasma-free or a medium with commercially available serum. Keller and Zam (1990) addressed the modification of gloquid culture for the first time, demonstrating that other sera can be substituted for fish plasma, with horse serum producing the best results.

However, work carried out by Uthaiwan *et al.* (2001, 2002) confirmed a greater affinity of glochidia for fish plasma when compared to a medium containing horse serum. The use of freeze-dried extracts of the fish *A. altparanae in* this study demonstrated that *in vitro* cultivation is possible *in a* simple, economical and efficient manner.

In the artificial culture technique, the gloquidia are removed, as described by Hudson and Isom (1984), or preferably by using a syringe with water to remove the larvae from the marsupium (Uthaiwan *et al.*, 2001, 2002; Lima *et al.*, 2006), but in this study, it was preferred to opt for the natural release of the gloquidia to avoid using unripe gloquidia. However, unlike previous publications, where the gloquidia are washed three to four times in autoclaved river water or reconstituted water and a final wash with Ringer's solution for Unionideo or a Hank's solution (Hudson and Isom, 1984; Milam *et al.* 2000), the gloquidia were washed repeatedly with distilled water and a final wash with a solution of antibiotics, and then placed to incubate in culture medium.

The original medium was composed of an Eagles solution with essential and non-essential amino acids in a Ringer's solution for Unionideo containing NaHCO3 for pH control, vitamins, antibiotics and glucose as the artificial part, and fish plasma as a natural protein source, in a final proportion of two thirds artificial medium and one third plasma (Isom and Hudson, 1982). As mentioned above, other media were tested by Hudson and Shelbourne (1990), and variations of the following specific components of the artificial medium were tested:

Ionic balance: initially, Isom and Hudson (1982) used a modification of Ringers for Unionideo described by Ellis *et al.* (1930), but tests have shown that preparations of balanced saline solutions such as Earle's or Hank's solution (Sigma) are useful for washing gloquids, as well as for their artificial transformation, although the yield may be slightly lower.

Protein source: fish plasma was used by Isom and Hudson (1982) as a protein additive. Hudson and Shelbourne (1990) working with *U. imbecillis* showed that rabbit serum can also be used as a protein source. They also demonstrated that the performance of rabbit serum is better than that of pig, horse, sheep, chicken, cattle and fetal serum. These results differ from the findings of Keller and Zam (1990) who reported that horse serum was more efficient as a protein additive. The combination of the aforementioned serums and plasmas was also tested by Hudson and Shelbourne (1990), who suggest using a combination of fish plasma/rabbit serum which normally promotes development equal to, or better than, fish plasma alone. Since some laboratories do not have access to fish plasma, and since infection rates are lower when using sterile serum obtained from biochemical supply companies, rabbit serum was considered the best alternative by Farris and Van Hassel (2006). In one case, *Elliptio angustata* gloquidia cultured in rabbit serum significantly outperformed fish plasma in transformation rates (Hudson and Shelbourne 1990). Although the values found in this study using freeze-dried fish extract were lower than those found in the literature, they were much higher than the metamorphosis values obtained by infesting the host fish. In addition, the preparation and preservation of the extract is much simpler and does not require fish to be kept in the laboratory. Antibiotics / Antimycotics: several washes of the gloquids removed from the marsupium, as described above, are essential to reduce the rate of infection by bacteria and fungi. Also helping to ensure a low infection rate are antibiotics (e.g. penicillin and streptomycin) and antimycotics (e.g. amphotericin B). These are normally found in most cell cultures and were the first to be used in the development of media for the maintenance and metamorphosis of glochidia (Isom and Hudson, 1982). Improvements to these techniques were made by isolating and growing bacterial cultures from gloquids and measuring the zones of inhibition of around a dozen commercially available antibiotics (Hudson and Isom, 1984). The three antibiotics (i.e. carbenicillin, gentamicin sulphate and rifampicin) with the best inhibitory effects were reported by Isom and Hudson (1982) and are currently the ones used in all laboratories working with glochidia cultures. Much later, bacteria that were isolated, by swabbing, from the gills *of Amblema plicata, P. catarata*, and *U. Imbecillis* were identified and measured for inhibitory effects (e.g. growth) by fifteen antibiotics (Loveless *et al.*, 1999). The assumption of this work is that the contamination of the artificial culture of gloquids probably comes from the gill tissue of the father, which houses these gloquids. The most effective antibiotics against the dozens of bacteria isolated were neomycin, ciprofloxacin and polymyxin B. Since the effect of these has never been evaluated on the success of metamorphosis, a mixture of antibiotics, including two of these new antibiotics was tested against a standard antibiotic control on the effect on the transformation of *U. Imbecillis* (Hudson and Isom, 1984). An advantage of using this increased number of antibiotics would be that a broad spectrum of bacteria can be better controlled than with the use of the antibiotics originally described by Isom and Hudson (1982). Invitrogen's antimycotic / fungicidal solution (table 3) did not contain the full range of antibiotics mentioned by the previous authors, which may have been one of the contamination factors during the incubation of the gloquids (figure 24).

Other components of the medium: most of the other components of the medium remained the same as first described by Isom and Hudson (1982). The essential and non-essential amino acids from Eagles' solution are mixed with Unionideo Ringer's with the addition of taurine and ornithine. Cultures seem to go well without the addition of these last two amino acids, but since they are found in fish plasma and since some species of shellfish may require these two amino acids, they are still used in most cultures. L-glutamine should be added weekly, due to its inability to remain stable in solution. Other culture media (e.g. M 199) have also been successful in transformation, even if their composition varies slightly from Eagles' modified medium (Isom and Hudson, 1982, Keller and Zam, 1990).

Artificially developed gloquids seem to have a lower lipid content than developing ones attached to fish (Fisher and Dimock, 2002). Tankersley (2000) showed that lipid levels in gloquids and juveniles vary according to each spawning and are influenced by the bivalve-mae. In addition, he indicated that the type of culture medium influenced the lipid content of gloquids and the development of subsequent juveniles. Fisher (2002) showed that gloquidia developed in artificial medium, with rabbit serum as a protein source, had lower lipid levels than those metamorphosed in the host fish. Hudson and Isom (1984) compared transformation rates with media containing squid oil (Artemate, Argent Chemical Laboratories) and cod liver oil. Although rates varied, cod liver oil seemed to be better than squid oil and also resulted in lower rates of fungal infection. The addition of this oil is thought to increase the lipid storage of glochidia (Tankersley, 2000).

Using the freeze-dried extract of *A. altparanae*, the survival and metamorphosis rates for *D. rotundus gratus* were 40% and 20% respectively. For *D. expansus* var. 2, the survival rate was 50% with 15% of individuals metamorphosing. Isom and Hudson (1982) and Hudson and Shelbourne (1990) showed that the transformation rate of glochidia with the inclusion of fish plasma ranged from 50 to 78 % of metamorphosed individuals. Uthaiwan et al. (2001, 2002, 2003) developed a simple technique for culturing gloquidia in artificial media, which resulted in a high metamorphosis rate (85 to 100 %), working with *H. myersiana*. Similar results were obtained with the European species *A. cygnea*, which showed a larval survival rate of 34.3%, while the proportion of larvae that underwent metamorphosis was 60.8% (Lima *et al.*, 2006).

Even though the survival and metamorphosis values obtained by previous authors were not achieved, the use of freeze-dried *A. altparanae* extract as a protein source in the culture medium provided the necessary components for the development of the gloquidia of the species tested. Although we do not have the composition of essential and non-essential amino acids of *A. altparanae*, only the body composition, 13.45 % protein and 6.05 % fat (Cotan *et al.*, 2006), our results suggest an efficiency in post-metamorphic survival, with 75% survival after 30 days. Authors such as Hudson and Isom (1984) were unable to maintain post-metamorphic juveniles for more than a week when working with the species *L. ovata*, *Fusconaia ebena*, *Ligumia recta*, *Pleurobema cordatum* and *Carunculina moesta*. Uthaiwan *et al.* (2001) using M 199 culture medium with *Oreochromis niloticus* plasma and horse serum, failed to maintain post-metamorphic juveniles of *H. myersiana* for more than 10 days. Uthaiwan *et al.* (2002) compared the effect of plasma from various fish species, finding post-metamorphic survival values of 2 months, 1 month, 2-3 weeks and 1-2 weeks with common carp, Nile tilapia and hybrid catfish, horse serum and striped catfish, respectively. Studies by Lima *et al* (2006) using *A. cygnea*, cultured in M 199 medium with *C. carpio* plasma, achieved post-metamorphic survival of 15 days.

A study by Hudson and Isom (1984) was perhaps the first successful attempt to culture juvenile freshwater bivalves in a laboratory. Since then, efforts by Buddensiek (1995), Gatenby *et al.* (1997), O'Beirn *et al.* (1998), Hanlon (2000), Jones *et al.* (2005) and other unpublished studies have diversified and improved culture techniques.

Gatenby *et al.* (1997) used Petri dishes with air to culture juveniles *of Villosa iris*, while O'Beirn *et al.* (1998) used recirculation systems for studies with *L. fasciola*. The juveniles of the genus Diplodon showed good survival when kept in Petri dishes and only after 15 days were they transferred to larger flasks with aeration.

These results are probably due to the podal feeding that juveniles experience shortly after metamorphosis (Yager *et al.*, 1994), and the reduced stress caused by air bubbles (Gatenby *et al.*, 1997). Jones *et al.* (2005) designed a "mesocosm" recirculation system for a variety of bivalve species. The mesocosm is a recirculation system with stages incorporating various natural biological processes, including the growth of bacteria and the decomposition of detritus. This mesocosm approach attempts to mimic a natural stream and provides a more diverse diet for the newly metamorphosed juveniles. All of these systems, however, require frequent maintenance and food supplementation with cultivated algae.

Juveniles *of U. imbecillis* and *Epioblasma triquetra* were successfully cultured in a medium with river water exposed to the sun for 1-4 days to increase the concentration of algae. In addition, the addition of silt improved the growth of juveniles in both species, while artificially fed, mixed culture of three species of algae resulted in starvation (Hudson and Isom, 1984). An observation supported by studies carried out by Yeager *et al.* (1994) suggests that post-metamorphic juveniles feed mainly on interstitial water from the sediment, through the foot furrow, so that silt and associated micro-debris and bacteria may be their main food source.

However, in experiments with river sediments, juveniles show a survival rate of 50-60% compared to juveniles kept in a simple, static culture for 90 days (Hudson and Roberts, 1997, Hudson and McKissic, 1999). Beck (2001) and Gatenby *et al.* (1997) demonstrated the importance of sediment in the culture of juvenile bivalves. The same was found for juveniles of the genus Diplodon. However, the use of sediment only proved to be effective when used with 12-day-old juveniles (Table 16).

The cause of the greatest mortality in the rearing of juveniles in this study was due to predation by

Turbellaria and Protists. Other authors cite predation as a cause of rearing difficulties. The average predation rate of *Macrostomum tuba* (Platyhelminthes, Turbellaria) on bivalve juveniles (number of juveniles consumed per worm per hour) was 0.26 ± 0.17 when the substrate was absent and 0.43 ± 0.18 when the substrate was present (Delp, 2002). Barnhart *et al.* (2008) reported that both *Macrostomum* and *Microstomum* ingest juvenile unionids. Henley *et al.* (2001) described *M. tuba* as a problem at the Virginia Polytechnic Institute.

Techniques that include the use of host fish (*in vivo*) and *in vitro* methods have generated viable juveniles for various purposes, including toxicity testing, *in situ* monitoring, as well as helping to stabilize declining populations through recovery plans for degraded species and habitats (Jacobson *et al.*, 1993; Yeager *et al.*, 1994).

The generation of juveniles can also help in the taxonomic identification of species. According to Bonetto (1961a), in addition to knowing the conquiliology and morphological characters of the adult, the ideal way to study the systematics of the genus *Diplodon* is to know the morphological characteristics of the gloquids, as these are always preserved whatever the geographical distribution and environment in which the species live.

Glochidial characteristics help in the identification and classification of this group of molluscs in which juvenile or adult individuals often do not provide interspecific diagnostic criteria (Mansur and Silva, 1999).

A characteristic of taxonomic value, cited by Bonetto (1959b) and Mansur and Campos- Velho (1990), would be the morphometry of the gloquidial shell.

The values for all the parameters measured in the three species studied are very close to those found by Mansur and Silva (1999) for five species of Diplodon: *D. martensi*, *D. berthae*, *D. koseritzi*, *D. ihering* and *D. charruanus*. The species *D. expansus* var. 1 and *D. expansus* var. 2 showed similar length variation to *D. ihering*, with 260 µm to 300 µm in length. With regard to height, *D. berthae*, with a range from 210 µm to 250 µm, was the species most similar to those in this study.

The displacements of the ventral tip of *D. expansus* var. 1 and *D. expansus* var. 2 are smaller than the values found by Mansur and Silva (1999) in the other species of Diplodon, while the angles of obliquity are larger. The glochidium *of D. rotundus gratus* was the largest among the three species studied, except for the angle of obliquity. Using the ANOVA test, it was possible to infer that *D. rotundus gratus* showed a significant difference compared to *D. expansus* var. 1 and *D. expansus* var. 2 for $p< 0.01$. However, the obliquity angles *of D. expansus* var. 1 and *D. expansus* var. 2 are not significantly different. Mansur (1999) did not carry out statistical tests for the morphometric measurements of the larvae found in the Diplodon species he studied.

This difficulty in identifying species of the genus Diplodon was evident in the results observed in this study, in the case of *D. expansus* (figure 17). However, the discussion of the great plasticity of the *D. expansus* species goes back further.

The description of *D. expansus* by Kuster (1856), made entirely from the shell, is very similar to the descriptions made by Ihering (1893) for *D. greeffeanus*. The type locality of Kuster's description was the Conigo River, Nova Friburgo, while the species described by Ihering comes from the Piracicaba River, Sao Paulo.

Ortmann (1921) made a series of corrections to the identification of South American naiad species, and using macroscopic anatomy (gills, palps, mantle), he divided the *D. expansus* group into *D. expansus*, *D. mogymirim* and *D. greeffeanus*. In this way, he describes the species *D. mogymirim*, with shell characteristics very similar to *D. greeffeanus*, but with different anatomy of the openings, gills and palps. The exhalant opening is slit-like, closed at the top; the closed portion is no more than twice the open portion. Exhalant opening as long as the inhalant, separated from it by a solid mantle connection. Inhalant opening with distinct papillae. Palpus moderate, subtriangular, posterior margin connected for about a third of its length. Gills somewhat wide, the outer gill subtriangular, slightly wider than the inner gill posteriorly; the inner gill subtrapezoidal, wider than the outer gill anteriorly, its anterior portion immediately behind the palpus. The inner lamella of the internal gill is completely connected to the abdominal sac.

Simone (2006), following the proposal made by Haas (1945) and Bonetto (1965), groups *D. mogymirim* and *D. greeffeanus* under *D. expansus*, as synonyms. However, as pointed out by Bonetto (1965), Haas distinguished this species by including a group of very heterogeneous forms. These conclusions were based on conquiliological characteristics, usually on the umbonal sculpture.

The three groups of *D. expansus* showed very distinct morphological characteristics of the gills (figure 20), but the small number of animals dissected meant that no conclusions could be drawn.

Conclusion

Based on the results obtained, it was possible to establish a methodology for the reproduction and cultivation of bivalves of the genus Diplodon. The methodology is summarized as follows:

- Washing the naturally eliminated gloquidia with distilled water and a final rinse with an antibiotic/antimycotic solution.

- Incubation of the gloquidia (100-150 gloquidia/plate) in Petri dishes (60x15 mm), with an artificial medium composed of two thirds of M 199 culture medium and one third of a solution of lyophilized fish extract as a protein source, plus the addition of antibiotics and antimycotics.

- The newly metamorphosed gloquids are kept in Petri dishes with autoclaved mine water, previously shaken with air diffusers

- On the fourth day of cultivation, food is added.

- On the twelfth day, the juveniles are transferred to Petri dishes with a 1 mm layer of sandy silt.

- On the twentieth day, they are placed in beakers with 1 mm of sediment and air diffusers to keep the microalgae in the water column and oxygenate the medium.

Despite the satisfactory results in terms of bivalve reproduction, hatching and metamorphosis of gloquidia and rearing juveniles of the limnetic bivalve species from the Mogi Guaçu River, the need for more in-depth taxonomic studies is clear for a restocking plan. And precisely because of the lack of such studies, any reintroduction program should be concerned with using animals from its own watershed.

Bibliographical references

Adams, H; Adams, A. 1858. The genera of recent Mollusca, arranged according to their organization. Volume 2. John van Voorst, London.

Ahlstedt, SA. 1981. The molluscan fauna of Copper Creek (Clinch River System) in southwestern Virginia. Bulletin of the American Malacological Union, 1: 4-6.

Almeida, ALAA. 2006. Temporal comparison of an association of limnetic bivalves from the Pardo River, municipality of Ribeirao Preto, State of Sao Paulo, Brazil. Master's dissertation, Ribeirao Preto School of Philosophy, Sciences and Letters/USP. 108 p.

Arey, LB. 1932. The nutrition of glochidia during metamorphosis. Journal of Morphology, 53 (1): 201-221.

Assad, LT; Bursztyn, M. 2000. Sustainable aquaculture. In Valenti, W.C. (ed). Aquaculture in Brazil: bases for sustainable development. CNPq/MCT, Brasilia, pp. 33-72.

Avelar, WEP. 1993. Functional anatomy of *Fossula fossiculifera* (D'Orbigny, 1843) (Bivalvia, Mycetopodidae). American Malacological Bulletin, 10(2): 129-138.

Avelar, WEP; Santos, SCD. 1991. Functional anatomy of *Castalia undosa undosa* (Martens, 1827) (Bivalvia: Hyriidae). The Veliger 34(1): 21-31.

Avelar, WEP; Mendonça, SHST. 1998. Aspects of gametogenesis of *Diplodon rotundus gratus* (Wagner, 1827) (Bivalvia: Hyriidae) in Brazil. American Malacological Bulletin, 14(2): 157-163.

Avelar, WEP; Cunha, AD. 2009. The anatomy and functional morphology of *Diplodon rhombeus fontainianus* (Orbigny, 1835) (Mollusca, Bivalvia, Hyriidae). Brazilian Journal of Biology, 69(4): 1153-1163.

Barfield, ML; Clem, SA; Farris, JL. 1997. Comparison of acute testing methods of *Utterbackia imbecillis* (Say) with variation on age, diet, and in vitro transformation method. Conservation and Management of Freshwater Mussels II: Initiatives for the Future. Proceedings of a UMRCC Symposium, St. Louis, USA, 293 p.

Barnhart, MC; Haag, WR; Roston, WN. 2008. Adaptations to host infection and larval parasitism in Unionoida. Journal of North American Benthological Society, 27(2): 230-394.

Bauer, G; Vogel, C. 1987. The parasitic stage of the freshwater pearl mussel (*Margaritifera margaritifera* L.). I: Host response to glochidiosis. Archiv für Hydrobiologie, 76(4): 393-402.

Beasley, CR. 2001. The impact of exploitation on freshwater mussels (Bivalvia: Hyriidae) in the Tocantins River, Brazil. Studies on Neotropical Fauna and Environment, 36(2): 159-165.

Beasley, CR; Tùry, E; Vale, WG; Tagliaro, CH. 2000. Reproductive cycle, management and conservation of *Paxyodon syrmatophorus* (Bivalvia: Hyriidae) from the Tocantins River, Brazil. Journal of Molluscan Studies, 66: 393-402.

Beasley, CR; Miranda, LQ; Alves, STM; Tagliaro, CH. 2003. Evaluation of methods for assessing brood size in freshwater mussel (Hyriidae). Journal of Molluscan Studies, 69: 159161.

Beck, KM. 2001. Development of an algal diet for rearing juvenile freshwater mussels (Unionidae). Master's thesis, Virginia Polytechnic Institute and State University, Virginia, USA. 58 p.

Bigham, SE. 2002. Host specificity of freshwater mussels: a critical factor in conservation. Master's thesis, Southwest Missouri State University, USA. 48 p.

Bogan, AE. 2008. Global diversity of freshwater mussels (Mollusca, Bivalvia) in freshwater. Hydrobiologia, 595: 139-147.

Bogan, AE; Hoeh, WR. 2000. On becoming cemented: evolutionary relationships among the genera in the freshwater bivalve family Etheriidae (Bivalvia: Unionoida). In Harper, EM; Taylor, JD; Crame, JA (eds), The Evolutionary Biology of the Bivalvia. Geological Society, London Special Publication, 177: 159-168.

Bonetto, AA. 1951. About the larval forms of Mutelidae ORTMANN. Jornadas Icticas, 1: 1-8.

Bonetto, AA. 1954. Nàyades of the Paranà River. El gènero Diplodon en el biotopo isleno del Paranà mèdio e inferior. Publicación Tècnica, Secretaria de Agricultura, Ganaderia e Industrias (Santa Fe), 62: 1-56.

Bonetto, AA. 1959a. Sobre algunas nuevas formas larvales de Hyriinae Ortmann. Actas del I Congreso Sudamericano de Zoologia, La Plata, 2: 33-41.

Bonetto, AA. 1959b. Contribución al conocimiento de las glochidias del genero Diplodon y su aplicación a los estudios sistemàticos. Actas del I Congreso Sudamericano de Zoologia, La Plata, 2: 43-59.

Bonetto, AA. 1960a. On the geographical distribution of natives in the Argentine Republic. Physis, T. XXII, n° . 63 (Sesiones Cient. Zoologia, Tucumàn), Argentina.

Bonetto, AA. 1960b. Sobre algunas formas larvales de Hyriinae Ortmann, 1° Congreso Sudamericano Zoologia, La Plata, 2.

Bonetto, AA. 1961a. Investigaciones acerca de las formas larvales en el gènero *Diplodon* y su aplicación a los estudios sistemàticos. Dirección General de Recursos Naturales, Santa Fé, Argentina, 48 p.

Bonetto, AA. 1961b. Nuevas notas sobre formas larvales de nàyades Sud y Centroamericanas. Physis, 21(62): 332-335.

Bonetto, AA. 1961c. Acerca de la distribución geogràfica de las nàyades en la Repûblica Argentina. Physis, 22(63): 259-268.

Bonetto, AA. 1962. Notes on *Diplodon charruanus* (Orb.) and *Diplodon rhuacoicus* (Orb.). Dirección General de Recursos Naturales, Santa Fé, Argentina, 10: 35-44.

Bonetto, AA. 1964. Las especies do genero Diplodon (Moll. Unionacea) en los rios de la pendiente Atlàntica del Sur de Brasil. Physis, 68: 323-328.

Bonetto, AA. 1965. Species of the genus Diplodon in the Rio de la Plata river system. Proceedings of the II Latin American Congress of Zoology, Sao Paulo, 1962, vol. 2: 37-54.

Bonetto, AA; Ezcurra, ID. 1963. Notas Malacológicas I. El desarrollo del glochidium *de Diplodon delodontus delodontus* (Lam.), Physis, B. Aires, 24(67): 17- 21.

Bonetto, AA; Maciel, I; Pignalberi, C. 1962. Algunos factores ecológicos vinculados a la distribución geogràfica de las nàyades en el rio Paranà y sus afluentes. Dirección General de Recursos Naturales Santa Fé, Argentina, 12: 167-175.

Boss, KJ. 1982. Mollusca. In S.P. Parker (ed.). Synopsis and Classification of Living Organisms. Volume 1. McGraw-Hill Book Company, New York. pp. 945-1166.

Bradshaw, M. 1984. Permian non-marine bivalves from the Ohio Range, Antarctica. Alcheringa, 8: 305-309.

Brusca, RC; Brusca, GJ. 2007. Invertebrates. Guanabara Koogan, Sao Paulo, Brazil. p. 1092.

Buddensiek, V. 1995. The culture of juvenile freshwater pearl *mussels*

Margaritifera margaritifera L. In cages: a contribution to conservation programs and the knowledge of habitat requirements. Biological Conservation, 74: 33-40.

Christian, AD; Smith, BN; Berg, DJ; Smoot, JC; Findlay, RH. 2004. Trophic position and potential food source of 2 species of unionid bivalves (Mollusca: Unionidae) in 2 small Ohio streams. Journal of North American Benthological Society, 23(1): 101-113.

Clem, SA. 1998. Complexities in testing early lifestage responses of freshwater mussels to metals and contaminated sediments. Master's thesis. Arkansas State University. 89 p.

Coker, RE; Shira, AF; Clark, HW; Howard, AD. 1921. Natural history and propagation of freshwater mussels. Bulletin of the Bureau of Fisheries, 37: 77-81.

Cope, OB. 1959. New parasite records from stickleback and salmon in an Alaskan stream. Transaction of American Microscopy Society, 789: 157162.

Cosgrove, PJ; Hastie, LC. 2001. Conservation of threatened freshwater pearl mussel populations: river management, mussel translocation and conflict resolution. Biological Conservation, 99: 183-190.

Cotan, J.L.V., Lanna, E.A.T., Bomfim, M.A.D., Donzele, J.L., Ribeiro, F.B. and Serafini, M.A. 2006. Levels of digestive energy and crude protein in diets for lambari tambiù fry. Revista Brasileira de Zootecnia, 35(3), 634-640.

Davis, GM; Fuller, SLH. 1981. Genetic relationships among Recent Unionacea (Bivalvia) of North America. Malacologia, 20: 217-253.

Delp, AM. 2002. Flatworm predation on juvenile freshwater mussels. Master's thesis. Southwest Missouri State University. 31 p.

Dodd, BJ; Barnhart, MC; Rogers-Lowery, CL; Fobian, TB; Dimock, RV. 2005. Cross-resistance of largemouth bass to glochidia of unionid mussels. Journal of Parasitology, 91(5): 1064-1072.

Dodd, BJ; Barnhart, MC; Rogers-Lowery, CL; Fobian, TB; Dimock, RV. 2006. Persistence of host response against glochidia larvae in *Micropterus salmoides*. Fish and Shellfish Immunology, 21(5): 473-484.

Dudgeon, D; Morton, B. 1984. Site selection and attachment duration of *Anodonta woodiana* (Bivalvia: Unionacea) glochidia on fish hosts. Journal of Zoology of London, 204: 355-362.

Ellis, MM; Ellis, MD. 1926. Growth and transformation of parasitic glochidias in physiological nutrient solutions. Science, 54(1667): 579-580.

Ellis MM; Merrick, AD; Ellis, MD. 1930. The blood of North American freshwater mussels under normal and adverse conditions. Bulletin of Bureau Fisheries, 56: 509-542.

Farris, J.L. and Van Hessel, J.H. 2006. Freshwater bivalve ecotoxicology. SETAC Press, Florida.

Figueiredo, J; Hoorn, C; Van der Ven, P; Soares, E. 2009. Late miocene onset of the Amazon river and the Amazon deep-sea fan: Evidence from the Foz do Amazonas basin. Geology, 37(7): 619-622.

Fisher, G. 2002. Physiological condition of in vitro- and in vivo-reared juvenile mussels. Abstract. Freshwater Mussel Conservation Society, Shepherdstown, USA. 11 p.

Fisher, GR; Dimock, RV. 2002. Morphological and molecular changes during metamorphosis in *Utterbackia imbecillis* (Bivalvia: Unionidae). Journal of Molluscan Studies, 68(2): 159-164.

Fisher, GR; Dimock, RV. 2006. Indicators of physiological condition in juveniles of *Utterbackia imbecillis* (Bivalvia: Unionidae): A comparison of rearing techniques. American Malacological Bulletin, 21(1-2): 23-29.

Fitzgerald, P. 2002. Tectonics and landscape evolution of the Antartic plate since the breakup of Gondwana, with an emphasis on the West Antartic Rift System and the Transantartic Mountains. Royal Society of New Zealand Bulletin, 35: 453-469.

Frierson, LS. 1927. A Classified and Annotated Check List of the North American Naiades. Baylor University Press, Waco, Texas.

Fryer, G. 1954. Development in a mutelid Lamellibranch. Nature, 183: 13421343.

Fryer, G. 1961. The developmental history of *Mutela bourguignati* (Ancey) (Mollusca: Bivalvia). Philosophical Transactions of the Royal Society of London B, 244: 259-298.

Gatenby, CM; Parker, BC; Neves, RJ. 1997. Growth and survival of juvenile rainbow mussels, *Villosa iris* (Lea, 1829) (BIVALVIA: UNIONIDAE), reared on algal diets and sediment. American Malacological Bulletin, 14(1): 57-66.

Giribet, G; Wheeler, W. 2002. On bivalve phylogeny: a high-level analysis of the Bivalvia (Mollusca) based on combined morphology and DNA sequence data. Invertebrate Biology, 121(4): 271-324.

Giusti, F; Castagnolo, L; Moretti, F; Renzoni, L; Renzoni, A. 1975. The reproductive cycle and the glochidium of *Anodonta cygnea* from Lago Trasimento (Central Italy). Monitore Zoologico Italiano, 9: 99-118.

Graf, DL. 1997. Sympatric speciation of freshwater mussels (Bivalvia: Unionoidea): a model. American Malacological Bulletin, 14(1):35-40.

Graf, DL. 2000. The Etherioidea revisited: a phylogenetic analysis of Hyriid relationships (Mollusca: Bivalvia: Paleoheterodonta: Unionoida). Occasional Papers of the Museum of Zoology, University of Michigan, 729: 1-21.

Graf, DL. 2002. Molecular phylogenetics analysis of two problematic freshwater mussel genera (*UNIO* and *GONIDEA*) and a re-evaluation of classification of neartic UNIONIDAE (BIVALVIA: PALAEHETERODONTA: UNIONOIDA). Journal of Molluscan Studies, 68: 65-71.

Graf, DL. 2007. Palearctic freshwater mussel (Mollusca: Bivalvia: Unionoida) diversity and the Comparatory Method as a species concept. Proceedings of the Academy of Natural Sciences of Philadelphia, 156: 71-88.

Graf, DL; Ó'Foighil, D. 2000. The evolution of brooding characters among the freshwater pearly mussels (Bivalvia: Uniondoidea) of North America. Journal of Molluscan Studies, 66: 157-170.

Graf, DL; Cummings, KS. 2006. Palaeoheterodont diversity (Mollusca: Trigonioida + Unionoida): what we know and what we wish we knew about freshwater mussel evolution. Zoological Journal of the Linnean Society, 148: 343-394.

Graf, DL; Cummings, KS. 2007. Review of the systematic and global diversity of freshwater mussel species (Bivalvia: Unionoida). Journal of Molluscan Studies, 73: 291-314.

Gray, EV; Lellis, WA; Cole, JC; Johnson, CS. 2002. Host identification for *Strophitus undulatus* (Bivalvia: Unionidae), the creeper, in the Upper Susquehannan River basin, Pennsylvania. American Midland Naturalist, 147(1): 153-161.

Haag, WR; Warren, ML. 2003. Host fishes and infection strategies of freshwater mussels in large Mobile Basin streams, USA. Journal of the North American Benthological Society, 22(1): 78-91.

Haas, F. 1945. Some remarkable shells of a South American freshwater mussel. Fieldiana- Zoology, 31: 15-30.

Haas, F. 1969. Superfamily Unionacea. In R. Martens and W. Hennig (eds.). Das Tierrich, Lieferung 88, Walter de Gruyter and Co., Berlin. 663 pp.

Hanlon, SD. 2000. Release of juvenile mussels into a fish hatchery raceway: comparison of techniques. Master's thesis. Virginia Polytechnic and State University. 131 p.

Hastie, LC; Cosgrove, PJ; Ellis, N; Gaywood, M. 2003. The Threat of Climate Change to Freshwater Pearl Mussel Populations. Ambio, 32(1): 40-47.

Heard, WH; Hendrix, SS. 1964. Behavior of unionid glochidia. Annual Report of the American Malacological Union.

Heard, WH; Guckert, RH. 1971. A re-evaluation of the Recent Unionacea (Pelecypoda) of North America. Malacologia, 10: 333-355.

Hebling, NJ. 1976. The functional Morphology of *Anodontites trapezeus* (Spix) and *Anodontites trapesialis* (Lamarck) (Bivalvia: Mycetopodidae). Boletim de Zoologia da Universidade de Sao Paulo, 1: 265-298.

Hebling, NJ; Penteado, AMG. 1974. Functional anatomy *of Diplodon rotundus gratus* Wagner, 1827 (MOLLUSCA, BIVALVIA). Revista Brasileira de Biologia, 34(1): 67-80.

Henley, WF. 2002. Evaluation of diet, gametogenesis and hermaphroditism in freshwater mussels (BIVALVIA: UNIONIDAE). Doctoral dissertation. Virginia Polytechnic Institute and State University, Virginia, USA. 148 p.

Henley, WH; Neves, RJ. 1999. Recovery status of freshwater mussels (Bivalvia: Unionidae) in the North Fork Holston River, Virginia. American Malacological Bulletin, 15: 65-73.

Henley, W.; Neves, RJ. 2001. Behavioral responses of glochidia of freshwater mussels (BIVALVIA: UNIONIDAE) to chemical cues of fish. American Malacological Bulletin, 16(1/2): 131-135.

Henley, WF; Zimmerman, LL; Neves, RJ; Kidd, MR. 2001. Design and evaluation of recirculating water systems for maintenance and propagation of freshwater mussels. North American Journal of Aquaculture 63: 144-155.

Hoeh, WR; Bogan, AE; Heard, WH; Chapman, EG. 2009. Palaeoheterodont phylogeny, character evolution, diversity and phylogenetic classification: a reflection on methods of analysis. Malacologia 51: 307-317.

Hoggarth, MA. 1992. An examination of the glochidia-host relationships reported in the literature for North American species of Unionacea (Mollusca: Bivalvia). Malacology Data Net 3:1-30.

Hoggarth, MA. 1999. Descriptions of some of the glochidia of the Unionidae (Mollusca: Bivalvia). Malacologia, 41: 180-298.

Hove, M.C. and Neves, R.J. 1994. Life history of the endangered James spinymussel *Pleurobema collina* (Conrad, 1837) (Mollusca: Unionidae). American Malacological Bulletin, 11, 29-40.

Hove, MC; Hillegass, KR; Kurth, JE; Pepi, VE; Lee, CJ; Knudsen, KA; Kapuscinski, AR; Mahoney, PA; Bomier, MM. 2000. Considerations for conducting host suitability studies. Freshwater mollusk symposia proceedings, part I. Proceedings of the Conservation, Captive Care, and Propagation of Freshwater Mussels Symposium, Ohio Biological Survey, Columbus, 27-34.

Howard, A.D. 1915. Some exceptional cases of breeding among Unionidae. The Nautilus, 29: 4-11.

Hudson, RG; Isom, B. 1984. In vitro culture of parasitic freshwater mussel glochidia. The Nautilus,

96: 147-151.

Hudson, RG; Shelbourne, CW. 1990. Improved in vitro culture of parasitic freshwater mussel glochidia Report to the Tennessee Valley Authority.

Hudson, RG; Roberts, CE. 1997. Chronic toxicity effects of Pigeon River sediment when used to culture freshwater juvenile mussels. Report Tennessee Wildlife Resources Division, p. 61.

Hudson, RG; McKissic, RQ. 1999. Chronic toxicity effects of Conasauga and Teleco River sediments when used to culture freshwater juvenile mussels. Report Tennessee Wildlife Resources Division, p. 49.

Hudson, RG; Wade, DC; McKinney, AD; Martin, RD. 1994. Comparative response of selected non-target aquatic organisms to commercial molluscacide. Abstract. 15th Annual Meeting, Society of Environmental Toxicology and Chemistry, Denver, USA. 186 p.

Hudson, RG; Barfield, ML; McKinney, AD. 1996. Species specificity and effect of pH on the response of freshwater mussel juveniles to acute copper toxicity. 88th Annual National Shellfish Association Meeting, Baltimore, USA. 25 p.

Hupé, MH. 1857. Mollusks. In: Animaux Nouveaux ou Rares Recueillis Pendant Léxpedition dans lês Parties Centrales de Lâmerique du Sud, de Rio de Janeiro a Lima, et de Lima au Para. P. Bertrand, Libraire (editeur). Paris, 123 p.

Ihering, H. 1890. Revision der von Spix Brasilien gesammelten najaden. Archiv fur Naturgeschichte, 56(1): 117-170.

Ihering, R. 1893. Najaden von S.Paulo unb die geographische ver breitung der susswasserfauna von sudamerika. Archiv fur naturgeschichte, 59: 45140.

Isom, BG; Hudson, RG. 1982. Culture of freshwater mussel glochidia in an artificial habitat utilizing complex liquid growth media. US Patent 4,449,480. 18 pp.

Jacobson, PJ; Farris, JL; Cherry, DS; Neves, RJ. 1993. Juvenile freshwater mussel (Bivalvia: Unionidae) responses to acute toxicity testing with cooper. Environmental Toxicology and Chemistry, 12: 879-883.

Jenkinson, J. 1982. Cumberlandian mollusk conservation program, In A.C., Miller (ed), Report of freshwater mollusks workshop, US Army Engineer Waterways Experiment Station, Environmental Laboratory, Vicksburg, USA. 185 p.

Johnson, IS; Keller, AE; Zam, SG. 1993. A method for conducting acute toxicity tests with the early life stages of freshwater mussels. Standard Technical Publication 1179, ASTM, Philadelphia.

Jones, JW; Mair, RA; Neves, RJ. 2005. Factors Affecting Survival and Growth of Juvenile Freshwater Mussels Cultured in Recirculating Aquaculture Systems. North American Journal of Aquaculture, 67: 210-220.

Jones, JW; Hallerman, EM; Neves, RJ. 2006. Genetic management guidelines for captive propagation of freshwater mussels (Unionoidea). Journal of Shellfish Research, 25(2): 527535.

Kat, PW. 1984. Parasitism and the Unionacea (BIVALVIA). Biological Reviews, 59: 189-207.

Keller, AE; Zam, SG. 1990. Simplification of in vitro culture techniques for freshwater mussels. Environmental Toxicological Chemistry, 9: 1291- 1296.

Kirk, SG; Layzer, JB. 1997. Induced metamorphosis of freshwater mussel glochidia on nonhost fish. The Nautilus, 110(3): 102-106.

Kondo, Y; Sano, SI. 2009. Origination of extant heteroconch families: Ecological and environmental patterns in post-Paleozoic bivalve diversification. Paleontological Research, 13(1): 39-44.

Korniushin, AV. 1998. Review of the studies on freshwater mollusk systematic carried out by the Russian taxonomic school. Bivalvia I: Malacological Review Supplement, 7: 65-82.

Kuster, HC. 1856. Systematisches Conchylien Cabinet (1848), 1852-1862.

Lee, HH; Mora, DA. 2005. Survival of glochidia of freshwater mussels, *Pyganodon grandis* (Mollusca: Unionidae), *in vitro* in the United States. World Aquaculture, 36(4): 56-60.

Lefevre, G; Curtis, WC. 1912. Studies on the reproduction and artificial propagation of fresh-water mussels. Bulletin of the Bureau of Fisheries 30: 107-201.

Lima, P; Uthaiwan, K; Satit, K; Machado, J. 2006. In vitro culture of glochidia from the freshwater mussel *Anodonta cygnea*. Invertebrate Biology, 125(1): 34-44.

Loveless, R; Raymond, S; Inman, J; Hudson, R; Fairfax, M. 1999. Isolation, identification and control of bacteria which threaten in vitro cultures of freshwater mussel glochidias. 1st Symposium of the Freshwater Mollusk Conservation Society, Chattanooga. 79 p.

Mackie, GL. 1984. Bivalves. In Karl Wilbur (ed.) THE MOLLUSCA, vol 7, chap 5, pg. 351-418.

Mansur, MCD. 1970. List of bivalve molluscs of the Hyriidae and Mycetopodidae families for the state of Rio Grande do Sul. Iheringia, Zoologia, 39: 33-95.

Mansur, MCD. 1972. Morphology of the digestive system of *Castalia undosa martensi* (IHERING,

1891) (BIVALVIA, HYRIIDAE). Iheringia, Zoologia, 41: 21-34.

Mansur, MCD. 1973. Morphology of the digestive system of species of the genus Diplodon Spix, 1827 from the Guaiba River, Rio Grande do Sul (Unionacea, Hyriidae). Iheringia Zoologia, 43: 7590.

Mansur, MCD. 1999. Glochidium of *Diplodon martensi* (Ihering) (Mollusca, Bivalvia, Hyriidae) and its parasitic cycle. Revista Brasileira de Zoologia, 16(2): 185-194.

Mansur, MCD; Anflor, LM. 1981. Morphological differences between *Diplodon charruanus* ORBIGNY, 1835 and *D. Pilsbryi* MARSHALL, 1928 (Bivalvia, Hyriidae). Iheringia, Zoologia, 60: 101116.

Mansur, MCD; Campos-Velho, NMR. 1990. Techniques for the study of Hyriidae glochiids (Mollusca, Bivalvia, Unionoida). Acta Biològica Leopoldensia, Sao Leopoldo, 12(1): 518.

Mansur, MCD; Silva, MGO. 1999. Description of glochidia of five species of freshwater mussels (HYRIIDAE: UNIONOIDEA) from South America. Malacologia, 41(2): 475-483.

Mansur, MCD; Campos-Velho, NMR. 2000. The glochidium of *Castalia martensi* (Ihering 1891) (Bivalvia, Unionoida: Hyriidae). Heldia, 3(1): 06-10.

Mansur, MCD; Schulz, C; Garces, LMMP. 1987. Freshwater bivalve molluscs: identification of genera from southern and eastern Brazil. Acta Biológica Leopoldinense, 9(2): 181-202.

Mansur, MCD; Santos, CP; Darrigan, G; Heydrich, I; Callil, CT; Cardoso, FR. 2003. First qualitative and quantitative data of the golden mussel, *Limnoperna fortunei* (Dunker), in the Jacui Delta, Guaiba Lagoon and Patos Lagoon, Rio Grande do Sul, Brazil and some aspects of its invasion in the new environment. Revista Brasileira de Zoologia, 20(1): 75-84.

Martinez, S; Del Rio, C. 2002. The Miocene and Recent Malacological Provinces of the Southwestern Atlantic. Anales de Biologia, 24: 121-130.

Martinez, S; Figueiras, A; Da Silva, JS. 1993. A New Unionoidea (Mollusca, Bivalvia) from the Tacuarembó Formation (Upper Triassic-Upper Jurassic), Uruguay. Journal of Paleontology, 67(6): 962-965.

Mayr, E. 1963. Animal Species and Evolution. Belknap Press of Harvard University Press, Cambridge, Massachusetts.

Mayr, E. 1971. Populations, Species, and Evolution. Harvard University Press, second printing , 453 p.

McMahon, RF; Bogan, AE. 2001. Mollusca: Bivalvia. In: Ecology and Classification of North American

Freshwater Invertebrates. 2nd Edition, Academic Press. Pp. 331-429.

McMichael, DF; Hiscock, ID. 1958. A monograph of the freshwater mussels (MOLLUSCA: PELECYPODA) of the Australian region. Australian Journal of Marine and Freshwater Research, 9(3): 372-508.

Meyers, TR; Millimann, RE; Fustish, CA. 1980. Glochidiosis of salmonid fishes. IV. Humoral and tissue responses of coho and chinook salmon to experimental infection with *Margaritifera*. Journal of Parasitology, 66: 274-281.

Milam, CD; Farris, JL; Van Hassel, J; Barfield, ML. 2000. Reintroduction of native freshwater mussels using in vivo and in vitro propagation techniques. Freshwater mollusk symposia proceedings, part I. Proceedings of the Conservation, Captive Care, and Propagation of Freshwater Mussels Symposium, Ohio Biological Survey, Columbus, 53-61.

Modell, H. 1942. Das natürliche System der Najaden. Archiv für Molluskenkunde 74: 161191.

Modell, H. 1964. Das natürliche System der Najaden. Archiv für Molluskenkunde 93: 71129.

Morrison, JPE. 1956. Family relationships of the North American fresh water mussels. American Malacological Union Annual Reports for 1955: 16-17.

Morrison, JPE. 1973. The families of the pearly freshwater mussels. Bulletin of the American Malacological Union, Inc. 38: 45-46.

Morton, LS; Herbst, R. 2001. New species of the genus Diplodon Spix (Bivalvia, Unionidea) from the Middle Jurassic (Formación La Matilde), provincia de Santa Cruz, Argentina. Revista Museo Argentino Ciencias Naturales, 3(2): 159-164.

Mummert, A; Newcomb, TJ; Neves, RJ; Parker, B. 2006. Evaluation of a recirculating pond system for rearing juvenile freshwater mussels at White Sulphur Springs National Fish Hatchery, West Virginia, USA. American Malacological Bulletin, 21(1-2): 1-10.

Neves, FF. 2005. Prospective analysis of erosion risks in the Bonito River watershed (Descalvado - SP), potentially polluted by farm waste. Master's dissertation. Sao Carlos School of Engineering, USP, Sao Carlos, 199 p.

Neves, RJ. 1999. Conservation and commerce: management of freshwater mussel (BIVALVIA: UNIONOIDEA) resources in the United States. Malacologia, 41(2): 461-474.

Neves, RJ; Zale, AV. 1982. Freshwater mussels (Unionidae) of Big Moccasin creek, southwest Virginia. The Nautilus, 96(2): 52-54.

Neves, RJ; Weaver, LR; Zale, AV. 1985. An evaluation of host fish suitability for glochidia of *V. vanuxemi* and *V. nebulosa* (Pelecypoda: Unionidae). American Midland Naturalist, 113(1): 13-19.

Neves, RJ; Bogan, AE; Williams, JD; Ahlstedt, SA; Hartfield, PW. 1997. Status of aquatic mollusks in the southeastern United States: a downward spiral of diversity. In: G.W.Benz and D.E.Collins (ed) Aquatic fauna in peril: the Southeastern perspective, .Special Publication 1.

Decatur, GA, USA: Southeast Aquatic Research Institute, Lenz Design and Communications. p 43-85.

Newell, ND. 1969. Classification of Bivalvia. pp. N205-N224. In R.C. Moore (ed.) Treatise on Invertebrate Paleontology, Part N, Mollusca, 6. Volume 1: Bivalvia. Geological Society of America and University of Kansas Press, Lawrence, Kansas.

O'Beirn, FX; Neves, RJ; Steg, MB. 1998. Survival and growth of juvenile freshwater mussels (Unionidae) in a recirculating aquaculture system. American Malacological Bulletin, 14(2): 165-171.

O'Connel, MT; Neves, RJ. 1999. Evidence of immunological responses by a host fish (*Ambloplites* rupestris) and two non-host fishes (*Cyprinus carpio* and *Carassius auratus*) to glochidia of a freshwater mussel (*Villosa iris*). Journal of Freshwater Ecology, 14: 71-78.

Ortmann, A.E. 1910. A new system of the Unionidae. Nautilus, 23: 114-120

Ortmann, A.E. 1911. The anatomical structure of certain exotic Najades compared with that of the North American forms. Nautilus, 24: 103-108, 114-120, 127-131.

Ortmann, A.E. 1912. Notes upon the families and genera of the Najades. Annals of the Carnegie Museum, 8: 222-365.

Ortmann, AE. 1921. South American Naiades: a contribution to the knowledge of the freshwater mussels of South America. Memoirs of the Carnegie Museum, 8(3): 451-670.

Owen, C; Alexander, JE; McGregor, M. 2009. Control of microbial contamination during in vitro culture of larval unionid mussels. Invertebrate Reproduction Development, in press.

Pain, T; Woodward, FR. 1961. A revision of the freshwater mussels of the family Etheriidae. Journal of Conchology, 25: 2-8.

Parodiz, JJ. 1969. The Tertiary Non-Marine Mollusca of South America. Systematic Zoology, 19(3): 306-307.

Parodiz, JJ. 1973. The species complex of *Diplodon delodontus* (Lamarck) (Unionacea - Hyriidae). Malacologia, 14: 247-270.

Parodiz, JJ. 1982. Distribution and origin of the continental South America malacofauna. Malacologia, 22(1-2): 421-425.

Parodiz, JJ; Bonetto, AA. 1963. Taxonomy and zoogeographic relationships of the South American naiades (PELECYPODA: UNIONACEA: MUTELACEA). Malacologia, 1(2): 179-213.

Penn, GH. 1939. A study of the life cycle of the freshwater mussel, *Anodonta grandis*, in New Orleans. Nautilus, 52: 99-101.

Pillow, MJ; Crownhart, AK; Barnhart, MC. 2008. The role of conglutinates in the survival of glochidias larvae. 6th Biennial International Symposium of the Freshwater Mollusk Conservation Society, April 19 - 24, Baltimore, Maryland.

Pilsbry, HA. 1911. Non-marine mollusca of Patagonia. Reprinted from the Reports of Princeton University Expedition to Patagonia, 1896-1899, vol. III.

Rafinesque, CS. 1820. Monographie des coquilles bivalves fluviatiles de la Riviere Ohio, contenant douze genres et soixante-huit especes. Annales Générales des Sciences Physiques, Bruxelles 5(13): 287-322.

Ricciardi, A; Neves, RJ; Rasmussen, JB. 1998. Impending extinctions of North American freshwater mussels (Unionoida) following the zebra mussel (*Dreissena polymorpha*) invasion. Journal of Animal Ecology, 67(4): 613619.

Roe, KJ; Lydeard, C. 1998. Molecular systematics of the freshwater mussel genus *Potamilus* (Bivalvia: Unionidae). Malacologia, 39: 195-205.

Rogers, CL; Dimock, RV. 2003. Acquired resistance of bluegill sunfish *Lepomis macrochirus* to glochidia larvae of the freshwater mussel *Utterbackia imbecillis* (Bivalvia:Unionidae) following multiple infections. Journal of Parasitology, 89: 51-56.

Rogers-Lowery, CL; Dimock, RV. 2006. Encapsulation of attached ectoparasitic glochidia larvae of freshwater mussels by epithelial tissue on fins of naive and resistant host fish. Biological Bulletin, 210(1): 51-63

Rogers-Lowery, CL; Dimock, RV; Kuhn, RE. 2007. Antibody response of bluegill sunfish during development of acquired resistance against the larvae of the freshwater mussel *Utterbackia imbecillis*. Development and Comparative Immunology, 31: 143-155.

Rogers, SO. 1999. Population biology of the tan riffle shell (*Epioblasma florentina walkeri*) and the effects of substratum and light on juvenile mussel propagation. Master's thesis. Virginia Polytechnic

Institute and State University, Virginia, USA. 121 p.

Rosen, DE. 1978. Vicariance patterns and historical explanations in biogeography. Systematic Zoology, 27: 159-188.

Sachs, I. 1993. Transition strategies for the 21st century. In: Bursztyn, M. (org.) Para pensar o desenvolvimento sustentâvel. Ed. Brasiliense, Brasilia. Pp. 29-56.

Say, T. 1817. Conchology. In W. Nicholson (ed.), American Edition of the British Encyclopedia or Dictionary of Arts and Sciences, Comprising an Accurate and Popular View of the Present Improved State of Human Knowledge 2: 15 p.

Schneider, JA. 2001. Bivalve systematic during the 20thth century. Journal of Paleontology, 75(6): 1119-1127.

Scholz, H; Hartman, JH. 2007. Paleoenvironmental reconstruction of the upper cretaceous hell creek formation of the Williston basin, Montana, USA: implications from the quantitative analysis of unionid bivalve taxonomic diversity and morphologic disparity. Palaios, 22: 24-34.

Secretariat of the Convention on Biological Diversity. 2004. Solutions for sustainable mariculture - avoiding the adverse effects of mariculture on biological diversity. CBD Technical Series, 12.

Seshaiya, RV. 1969. Some observations on the life-histories of south Indian freshwater mussels. Malacologia, 9: 286-287.

Silva-Souza, AT; Eiras, JC. 2002 The Histopathology of the Infection *of Tilapia rendalli* and *Hypostomus regani* (Osteichthyes) by Lasidium Larvae *of Anodontites trapesialis* (Mollusca, Bivalvia) Memorias do Instituto Oswaldo Cruz, Rio de Janeiro, 97(3): 431-433.

Simone, LRL. 2006. Land and freshwater mollusks of Brazil. EGB, FAPESP, Sao Paulo.

Simpson, C.T. 1900. Synopsis of the Naiades, or pearly freshwater mussels. Proceedings of the. U.S. National Museum 22: 501-1044.

Simpson, C.T. 1914. A Descriptive Catalogue of the Naiades or Pearly Freshwater Mussels, Volumes I to III. Privately published by Bryant Walker, Detroit, Michigan.

Tankersley, RA. 2000. Fluorescence techniques for evaluating the lipid content of larval and juvenile mussels. Freshwater mollusk symposia proceedings, part I. Proceedings of the Conservation, Captive Care, and Propagation of Freshwater Mussels Symposium, Ohio Biological Survey, Columbus, 115-125.

Tankersley, RA; Butz, SW. 1998. Design, construction and evaluation of a laboratory-scale recirculating aquaculture system for the captive care of freshwater mussels. Freshwater mollusk symposia proceedings, part I. Proceedings of the Conservation, Captive Care, and Propagation of Freshwater Mussels Symposium, Ohio Biological Survey, Columbus, 127-134.

Telda, S; Fernando, CH. 1969. Observations on the glochidia of *Lampsilis radiata* (Gmelin) infesting Yellow perch, *Perca flavescens* (Mitchell) in the Bay of Quinte, Lake Ontario. Canadian Journal of Zoology, 47: 705-712.

Thiele, J. 1934. Handbuch der Systematischen Weichtierkunde. Verlag von Gustav Fisher, Berlin, 779-1154.

Tompa, AS. 1979. Life cycle completion of the freshwater clam *Lasmigona compressa* (Bivalvia: Unionidae) in an experimental host, *Lebistes reticulatus*. Veliger, 22: 188-190.

Uthaiwan, K; Noparatnaraporn, N; Machado, J. 2001. Culture of glochidia of the freshwater pearl mussel *Hyriopsis myersiana* (Lea, 1856) in artificial media. Aquaculture, 195: 61-69.

Uthaiwan, K; Pakkong, P; Noparatnaraporn, N; Vilarinho, L; Machado, J. 2002. Study of a suitable fish plasma for in vitro culture of glochidia *Hyriopsis myersiana*. Aquaculture, 209: 197-208.

Uthaiwan, K; Pakkong, P; Noparatnaraporn, N; Vilarinho, L; Machado, J. 2003. Studies on the plasma composition of fish hosts of the freshwater mussel, *Hyriopsis myersiana*, with implications for improvement of the medium for culture of glochidia. Invertebrate Reproduction and Development, 44(1): 53-61.

Villella, RF; Smith, DR; Lemarie, DP. 2004. Estimating Survival and Recruitment in a Freshwater Mussel Population Using Mark-recapture Techniques. American Midland Naturalist, 151:114-133.

Visconti, G; Elorriaga, EE; Parras, A. 2003. Marine sedimentites of the Barranca Final Formation (middle-superior Miocene of the Colorado basin), outcropping in the southeast of the province of La Pampa, Argentina. Revista de la Asociación Geológica Argentina, 58(2): 187193.

Wachtler, K; Mansur, MCD; Richter, T. 2001. Larval types and early postlarval biology in Naiads (Unionoida). In G. Bauer and K. Wachtler (eds) Ecology and evolution of the freshwater mussels Unionoida Ecological Studies, vol. 145, Springer Verlag, Berlin Heidelberg.

Waller, DL; Mitchell, LG. 1989. Gill tissue reactions in walleye *Stizostedion vitreum vitreum* and common carp *Cyprinus carpio* to glochidia of the freshwater mussel *Lampsilis radiata siliquoidea*. Diseases of Aquatic Organisms, 6: 81-87.

Watters, GT. 1996. Small dams as barriers to freshwater mussels (BIVALVIA: UNIONOIDEA) and their hosts. Biological Conservation, 75: 79-85.

Watters, GT. 2000. Freshwater mussels and water quality: a review of the effects of hydrologic and instream habitat alterations. Freshwater mollusk symposia proceedings, part I. Proceedings of the Conservation, Captive Care, and Propagation of Freshwater Mussels Symposium, Ohio Biological Survey, Columbus, 1-14.

Watters, GT; O'Dee, SH. 1996. Shedding of untransformed glochidia by fishes parasitized by *Lampsilis fasciola* Rafinesque, 1820 (Mollusca: Bivalvia: Unionidae): Evidence for acquired immunity in the field? Journal of Freshwater Ecology, 11: 383-389.

Watters, GT; O'Dee, SH. 1998. Metamorphosis of Freshwater Mussel Glochidia (Bivalvia: Unionidae) on Amphibians and Exotic Fishes. American Midland Naturalist, 139: 49-57.

Wesselingh, EP; Ranzi, A; Rasanen, ME. 2006. Miocene freshwater Mollusca from western Brazilian Amazonia. Scripta Geologica, 133: 419-437.

Williams, JD; Warren, LW; Cummings, KS; Harris. JL; Neves, RJ. 1993. Conservation status of freshwater mussels of the United States and Canada. Fisheries, 18 (9): 6-22.

Winterringer, R. 2004. Population dynamics and reproductive patterns of the federally endangered freshwater mussel, *Lampsilis Streckeri* (Frierson 1927). Master's thesis. Arkansas State University, Arkansas, USA. 74 p.

Yeager, BL, Neves, RJ. 1986. Reproductive cycle and fish hosts of the rabbit's foot mussel, *Quadrula cylindrical strigillata* (Mollusca: Unionidae) in the upper Tennessee River drainage. American Midland Naturalist, 116: 329-340.

Y eager, M. M., Cherry, D. S., Neves, R. J. 1994. Feeding and burrowing behaviors of juvenile rainbow mussels, *Villosa iris* (Bivalvia: Unionidae). Journal of the North American Benthological Society 13:217-222.

Y onge, CM. 1957. Mantle fusion in the Lamellibranchia. Pubblicazioni di Stazione Zoologica di Napoli, 29: 151-171.

Y onge, CM. 1982. Mantle margins with a revision of siphonal types in the Bivalvia. Journal of Molluscan Studies, 48: 102-103.

Y oung, M; Williams, J. 1984a. The reproductive biology of the freshwater pearl mussel *Margaritifera margaritifera* (LINN.) in Scotland. I. Field studies. Archives Für Hydrobiologia, 99(4):

405-422.

Young, M; Williams, J. 1984b. The reproductive biology of the freshwater pearl mussel *Margaritifera margaritifera* (LINN.) in Scotland. II. Laboratory studies. Archives Für Hydrobiologia, 100(1): 29-43.

Zale, AV; Neves, RJ. 1982. Fish hosts of four species of lampsiline mussels (MOLLUSCA: UNIONIDAE) in Big Moccasin Creek, Virginia. Canadian Journal of Zoology, 60: 2535-2542.

Zimmerman, LL. 2003. Propagation of Juvenile Freshwater Mussels (Bivalvia: Unionidae) and Assessment of Habitat Suitability for Restoration of Mussels in the Clinch River, Virginia. Master's thesis, Virginia Polytechnic Institute, Virginia, USA. 129 p.

Zimmerman, LL; Neves, RJ. 2002. Effects of temperature on duration of viability for glochidia of freshwater mussels (BIVALVIA: UNIONIDAE). American Malacological Bulletin, 17(1/2): 31-35.

Annex A. Current distribution of the Hyriidae family in Brazil, according to Simone (2006).

Superfamily	Family	Subfamily	Gender	Species	Occurrence
UNIONOIDEA	HYRIIDAE	HYRIINAE	Triplodon	*Triplodon chodo*	Amazon Basin, Guiana Basin
			Prisodon	*Prisodon corrugatus*	Amazon Basin
				Prisodon obliquus	Amazon Basin
			Paxyodon	*Paxyodon syrmatophorus*	Amazon Basin
			Callonaia	*Callonaia duprei*	Amazon Basin
			Castalia	*Castalia ambigua*	Amazon Basin, Tocantins, Sao Francisco
				Castalia inflata	Paranà Basin
				Castalia martensi	Lower Paranà Basin, Southern Brazil, Uruguay and Microbasins
					Atlantic
				Castalia multisulcata	Antilles, Venezuela, Guianas, Colombia, Acre, Microbasins
					Atlantic
				Castalia nehringi	Upper Paranà Basin
				Castalia orinocensis	Orinoco River
				Castalia psamoica	Paranà Basin
				Castalia undosa	Upper Paranà Basin
			Castaliella	*Castaliella sulcata*	Guyana, Suriname, Amazon Basin
			Diplodon	*Diplodon besckeanus*	Atlantic watersheds (BA to SP)
				Diplodon delodontus	Paranà Basin, Atlantic Microbasins (RS to Uruguay)
				Diplodon ellipticus	Upper San Francisco Basin, Upper Paranà Basin, Microbasins
					Atlantic (ES to RS)
				Diplodon expansus	Upper Paranà Basin, Atlantic Microbasins (RS to Uruguay)
				Diplodon fontainianus	Upper Paranà Basin, Atlantic Microbasins (ES to PR)
				Diplodon granosus	Amazon Basin, Paranà Basin, Atlantic Microbasins (BA
					to SP)
				Diplodon multistriatus	Amazon Basin, Paranà Basin, Atlantic Microbasins (BA
					to SC)
				Diplodon parallelipipedon	Amazon Basin, Paranà Basin
				Diplodon parodizi	Amazon Basin, Paranà Basin
				Diplodon patagonicus	Patagonia, Argentina
				Diplodon rhuacoicus	Amazon Basin, Paranà Basin, Atlantic Microbasins (RJ to
					Uruguay)
				Diplodon wymanii	Middle-Lower Paranà Basin
				Rhipidodonta *Rhipidodonta burroughiana*	Amazon Basin, Paranà Basin
				Rhipidodonta charruana	Paranà Basin, Atlantic Microbasins (SP to Uruguay)
				Rhipidodonta funebralis	Lower Paranà Basin
				Rhipidodonta grata	Paranà Basin
				Rhipidodonta	Amazon Basin, Paranà Basin

				hylaea	
				Rhipidodonta rhombea	Amazon Basin, San Francisco Basin
				Rhipidodonta suavidica	Amazon Basin, San Francisco Basin
				Rhipidodonta variabilis	Paranà Basin
				Diplodon rhuacoicus	Amazon Basin, Paranà Basin, Atlantic Microbasins (RJ to
					,, Uruguay)
				Diplodon wymanii	Middle-Lower Paranà Basin
				Rhipidodonta *Rhipidodonta burroughiana*	Amazon Basin, Paranà Basin
				Rhipidodonta charruana	Paranà Basin, Atlantic Microbasins (SP to Uruguay)
				Rhipidodonta funebralis	Lower Paranà Basin
				Rhipidodonta grata	Paranà Basin

Annex B. Abiotic parameters measured at the collection point on the Mogi Guaçu River, municipality of Porto Ferreira, Sao Paulo.

	2007							2008					
	Out	Nov	Dic	Jan	Feb	Sea	Apr	May	Jun	Jul	Aug	Set	Out
Turbidity (NTU)	15,9	50,5	50,5	167	147	157	87	38,9	32,4	28,4	37,1	24,1	19,3
Dissolved oxygen (g/ml)	6,2	6,9	7,3	7,6	8,2	7,9	7,2	9,5	9,2	8,8	9,1	9,1	10,2
Temperature (°C)	25,7	25,8	28	23	25	21,5	21,4	19	20,2	19,1	20,2	22,4	23,1
Electrical (US)	106,7	107,3	103,4	107,1	96,2	66,8	69,4	76,4	68,7	75,3	79,2	67,8	72,3
River flow (m3/s)	90,4	154,98	297,97	294,86	229,6	152,63	120,79	106,27	79,54	61,69	63,01	66,99	

Note: the river flow data is a monthly average from 1997 to 2007.

MIX
Papier aus verantwortungsvollen Quellen
Paper from responsible sources
FSC® C105338

Printed by Books on Demand GmbH, Norderstedt / Germany